VOLUME 7

COAL PROCESSING TECHNOLOGY

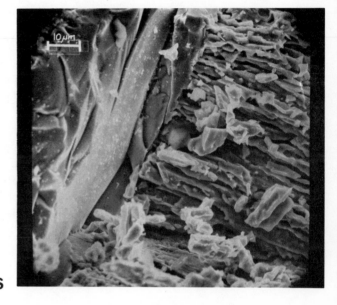

Prepared by Editors of

CHEMICAL ENGINEERING PROGRESS

a **CEP** technical manual published by
AMERICAN INSTITUTE OF CHEMICAL ENGINEERS
345 East 47 Street   New York, N.Y. 10017

TABLE OF CONTENTS

Development of a Process for the Supercritical Gas Extraction of Coal - J.C. Whitehead 1-8

Transient Model of a Moving-Bed Coal Gasifier - K.J. Daniel 9-11

Catalyzing Coal for Industrial Gasification - B.C. Kim, H.F. Feldman,
L.M. Curran, C.J. Robb 12-13

Operation of the Bi-Gas Coal Gasification Pilot Plant - J. Glenn 14-25

Extraction of Solvent - Refined Coal Liquids-E.G. Scheibel, A. Schneider, E.J. Hollstein 26-35

Operating Experience with the Westinghouse Gasifier - E.J. Chelen, D. Revay,
P. Cherish 36-42

Upgrading Solvent - Refined Coal by Hydrogenation-
J. Caspers, R.P. Van Driesen, K. Hastings, S. Morris 43-48

Hydrotreatment of COGAS Pyrolysis Oil via the H-Oil Process -
G. Nongbri, L. Lehman, L.I. Wisdom 49-63

Start-Up, Operation and Shutdown of a Wellman-Galusha Gasifer -
A.C. Mengon, J.S. Levin 64-69

Practical Operating Experience on a Riley Gasifer - W.P. Early,
R.A. Lisauskas, A.H. Rawdon 70-77

Hot Gas Cleaning for Advanced Coal Conversion Technologies -
D.F. Ciliberti, A.Y. Ranadive, D.L. Keairns, P.R. Mulik, M.A. Alvin 78-86

Desulfurization of Hot Coal-Derived Gas by Calcined Dolomite - E.J. Nemeth,
J.E. McGreal, J.C. Howell, J. Feinman 87-88

Gravimelt Process for Near Complete Chemical Removal -
R.A. Meyers, W.D. Hart, L.C. McClanathan 89-93

Sulfur Removal from Coal Char Using "Convert-Remove" Technology -
A.B. Tipton 94-97

A Large-Scale Cold Flow Scale-Up Test Facility - W.C. Yang,
S.S. Kim, J.A. Rylatt 98-107

Desulfurization of Iron Desulfide: Evaluation of Alternative Mechanisms -
D.P. Daugherty 108-123

Removal of Organic Sulfur by Low-Temperature Carbonization
of Illinois Coals - C.W. Kruse 124-134

Removal of Organic Sulfur from Coal - C.R. Porter, H.D. Kaesz, J.L. Leto,
J.J Giordano, W.R. Haas, E. Johnson, W.H. Berry, Jr. 135-140

Coal Desulfurization by Chlorinolysis—Phase II - J. Kalvinskas, W. Rohatgi 141-143

Coal Conversion R & D in Western Germany - R. Specks, A. Klusmann 144-154

Low Btu

C1108

Development of a Process for the Supercritical Gas Extraction of Coal

Thermobalance studies have shown that gas extraction residues are at least as reactive as the parent coal.

J.C. Whitehead

National Coal Board, Stoke Orchard
Cheltenham, England

The increasing uncertainty regarding the supply and cost of crude oil since the early 1970's has prompted an increase in coal lique- faction R & D in almost every country posses- sing significant reserves of coal. Although the UK is fortunate in having considerable reserves of oil and gas under the North Sea it is still predicted that towards the end of the century it will become increasingly necessary to supplement diminishing supplies with alter- native materials and forms of energy. Coal- based liquids have the potential to satisfy the demand for liquid transport fuels and chemical feedstocks and it is towards the production of these premium distillate products that coal liquefaction developments in the UK are being directed. Work at the Coal Research Establishment has led to the development of two separate liquefaction processes which are each based on initial research carried out in the mid-1960's. The processes are referred to as the Liquid Solvent Extraction Process (1) (LSE), and the Supercritical Gas Extraction Process (SGE).

Both processes are two-stage processes – an extraction stage in which coal extract is separated from mineral matter and undissolved coal and a subsequent extract hydrogenation stage – and as such have the important advan- tage of allowing optimisation of each stage, depending on coal feedstock and desired product slate, with regard to making efficient

use of the hydrogen consumed. Such consider- ations are extremely important in processes where high hydrogen content products, in relation to the coal feed, are being produced.

The SGE process, which is based on the ability of compressed gas to dissolve sig- nificant quantities of a high molecular sub- strate, represents a relatively novel approach to coal liquefaction. Although the solvent power of compressed gases was first demon- strated in the late nineteenth century it was not usefully employed to any extent until the middle of this century (2). Initial inves- tigations at the Coal Research Establishment, prompted by the work of Diepen and Scheffer (3), centred on the use of compressed gases, such as ethylene, to fractionate coal tar at low temperatures (<373 K); the potential of the technique was soon realised and efforts were soon redirected towards its application in a coal liquefaction process (4).

PROCESS PRINCIPLES

The basic principles which dictate the con- ditions under which a gas can act as an effec- tive solvent has been described in many pub- lications (2) and comment here will be limited to the implications of operating a coal lique- faction process under these desired con- ditions. Theory indicates that the solvent power of a gas increases with its density which, at a given pressure, will be a maximum at the minimum reduced temperature. The extracting gas is therefore chosen to have a critical temperature slightly below the temperature at which the extraction process is required to operate.

Experimental studies carried out on a wide range of binary systems have shown that significant quantities of a relatively high molecular weight substrate can be dissolved in a gas under conditions where:

(i) the substrate would exhibit a very low vapour pressure,

(ii) the gas possesses a density and viscosity both significantly lower than those of a liquid phase solvent.

These two factors have considerable significance since by judicious choice of the gaseous solvent:

(i) the liquids produced when coal is heated to temperatures in the range 673-723 K can be extracted without the need to raise the temperature to a level which could cause secondary repolymerisation reactions,

(ii) separation of undissolved residues, essential in a process where the coal extract is to be further treated using relatively sophisticated catalysts, is potentially less difficult than in liquid phase extraction processes,

(iii) separation of the coal extract and solvent is achieved simply and efficiently as a result of their widely different volatilities.

The preferred gases for use within a coal liquefaction scheme have critical temperatures in the range 573-673 K and are consequently liquids at ambient temperature. The basic operations within the process are shown in Figure 1. Coal is mixed with liquid phase solvent and pressurised to the extraction pressure. On heating the liquid phase solvent it passes through its critical temperature and expands to form a gas phase solvent. At the extraction temperature the coal is thermally depolymerised to form low molecular weight liquids, which are readily soluble in the gas phase, and a porous residue which contains virtually all the mineral matter associated with the coal. These residual solids are removed from the gas phase at extraction conditions, and the solids-free gas phase is depressurised to atmospheric pressure at which point the solvent power of the gas phase is reduced to an insignificant level and the coal liquids are precipitated from solution. Small amounts of gas and water generated during the extraction stage are removed after condensation of the solvent which is subsequently recycled.

The potential attractions of this process, pointed out previously, are to some extent offset by the fact that, since undissolved residue is separated from the gas phase at extraction conditions, a concentrated stream of residual solids must be withdrawn from the system via a separate route across a high pressure differential. This operation inevitably poses a number of mechanical engineering problems.

LABORATORY SCALE PROCESS DEVELOPMENT

Initial screening work (4,5) was carried out on a number of coals and solvents in mechanically stirred batch autoclave systems (10 dm^3 and 40 dm^3 capacity). This work identified toluene (Crit. Temp. 591 K) as a preferred solvent and showed that the yield of extract increased with increasing volatile matter content of the coal. Subsequently the bulk of experimental work has been carried out using low rank UK coals with volatile contents of the order of 40% daf. At this point in the development programme the maximum extract yield obtained was of the order of 20% by weight of dry ash free coal.

In order to overcome the inherent deficiencies of the batch autoclave systems being utilised e.g. long heat-up times, a small 'mini-bomb' system was constructed in order to carry out a detailed survey of the influence on extract yield of a number of operating variables. A flowsheet of this system is shown in Figure 2.

Using this equipment a fixed bed of coal (∼0.05 kg) could be rapidly heated to the extraction temperature and intimately contacted with a continuous flow of supercritical solvent. The coal bed was contained between sintered metal discs within a stainless steel tube (0.025 m i.d. x 0.3 m) which was supported on a metal frame together with a preheater coil and depressurising valve (V6). Solvent was supplied to the preheater via a high pressure metering pump. The whole device could be rapidly heated (2.5K.s^{-1}) to the extraction temperature by immersing it in an electrically heated air-fluidised sandbath.

Approximately 200 extractions were carried out to investigate the effect of the following variables.

(1) Extraction time

(2) Coal particle size

(3) Pressure

(4) Temperature

(5) Solvent to coal ratio

(6) Solvent type

(7) Solvent flowrate

An investigation of the effect of coal residence time, solvent flowrate and coal particle size on extract yield at 623 K showed that coal residence time was the variable which most influenced extract yield. In contrast, a five-fold increase in solvent flowrate had only a small effect on yield and a similar change in particle size was even less effective. It therefore appeared that the particles were readily penetrated by the supercritical solvent.

The influence of pressure and temperature on yield was substantial and it was shown that an extract yield of 35% by weight of daf coal could be obtained using toluene at 693 K, 27.5 MPa and a solvent to coal ratio of 5:1.

It was determined, that it is not essential for the extracting fluid to be above its critical temperature to be effective. For a given operating pressure the advantages of operating in a supercritical state are associated with the lower density and viscosity of the fluid compared with a sub-critical fluid.

Ultimate analyses of the recovered extracts, using toluene as solvent, showed little variation either with depth of extraction or extraction temperature but did illustrate the important fact that the hydrogen content was significantly higher than that of the parent coal. This ability of the solvent to extract a hydrogen-rich portion of the coal ultimately benefits the overall processing scheme by reducing the hydrogen requirement during second stage hydrocracking.

The results of these investigations, together with initial results from other related studies (Section 5.2) prompted the design and construction of two units which

operate with a continuous coal feed. One is a laboratory scale unit which processes 0.2 kgh^{-1} of coal and the other a self-contained plant processing 5 kgh^{-1} of coal.

PLANT SCALE DEVELOPMENT

The basic process engineering for the 5 kgh^{-1} plant was carried out by the Coal Research Establishment whilst detailed engineering and construction was performed by Woodall-Duckham Ltd. The total plant cost was of the order £750,000 and mechanical construction and testing was completed in September 1977. Preliminary process design showed there to be a lack of information regarding certain aspects of the gas/solid systems that would be present within the process and therefore a programme of laboratory scale model studies was carried out to provide information relevant to the design and operation of the plant. The main study involved an investigation of the fluidising properties of fine coal particles in a supercritical gas (6).

The plant is designed to withstand pressures up to 40 MPa and temperatures up to 773 K with the facility to feed up to 100 kgh^{-1} of liquid solvent. Coal feed size is dictated by plant pipework dimensions which in turn are dictated by plant throughput, consequently this plant utilises a nominal -210 μm coal feed. A flowsheet of the plant is given in Figure 3.

A coal-solvent slurry is fed from one of the two mix tanks (T_1, T_2) via centrifugal circulating pumps (P_1, P_2) to a high pressure metering pump (P_3). Clean solvent for plant start-up and flushing purposes is supplied from tank T_3. Pressurised slurry from P_3 is heated to extraction temperature in one of two heating coils immersed in electrically heated, air-fluidised sand baths (H_1, H_2) and the coal and solvent are contacted within a fluidised bed reactor (R). This vessel is 0.1 m i.d. and the solids residence time and solvent velocity can be adjusted by the use of internal sleeves which reduce the cross sectional area of the bed. Undissolved solids are transported away from the bed in the supercritical gas phase which contains dissolved coal extract. The solids are separated from the gas by a combination of separation devices housed within vessel S. Coal extract can be recovered from the supercritical solvent in two ways. By use of the high pressure coolers (C_1, C_2) the gas phase can be cooled and extract recovered from the depressurised condensate by distillation. Alternatively, the high pressure coolers may be by-passed and the gas phase depressurised; this causes the coal extract to be precipitated from solution, after which it is collected in T4 and the low pressure solvent vapour is condensed in the low pressure coolers (C3, C4) and collected in T5. There is no facility on the plant at this time for continuously removing solid residues; undissolved solids discharged from separator S are collected in the reservoir V and removed when the run is terminated.

Plant commissioning and the initial phase of the experimental programme have been carried out using Daw Mill Coal (Volatile Matter 40% daf) and toluene as solvent; this work comprised a total of twenty-six plant operations with approximately 30 kg of coal being processed during each operation. These operations have been designed to allow investigation of a range of operating conditions whilst also improving the performance and reliability of a number of operations and plant components. It is therefore convenient to consider the experimental programme in terms of plant investigations and process investigations.

Plant Investigations

At the termination of the contractors pre-commissioning operation there were a number of outstanding problems e.g. leaking pump glands, which, although they did not prevent plant operation, were the subject of immediate investigation. Commissioning of the plant was, in general, successful in that a relatively high number of operations were carried out to completion. There were no significant problems in the solids' transport regions of the plant and no evidence of particle agglomeration or deposition on the internal surfaces of pipework or vessels.

Slurry feeding. One major problem encountered during initial operations was that of overheating and subsequent failure of gland packings in the slurry recycle pumps P_1 and P_2. A number of different packing materials were tried, none of which gave satisfactory performance and consequently the pump glands were modified to accept Carbon/'NiResist' mechanical seals. These seals are flushed with clean fluid, recovered in a small hydrocyclone, from a portion of the pump discharge and have given approximately 1000 h trouble free operation to date.

Contacting. It has been shown that the weight of the contents of the fluidised bed can be reliably estimated via a differential pressure measurement across the bed. Using this

measurement it was established that, beyond the start-up period when the bed was filling, the weight of material in the bed continued to increase. Examination of the bed contents after approximately 6 h operation indicated, as expected, that the particle size distribution was considerably different to the coal feed. Size analysis and mineral matter content data shown in Table 1 illustrate the way in which large coal particles and heavy mineral matter particles accumulate in the bed, although the calculated size distribution of the whole residue is very similar to that of the coal feed.

In order to increase the effective operating time of the reactor, the system was prefed with the reactor contents from a previous operation and by this means it was shown that there was in fact no significant permanent accumulation of particles within the bed. Some studies have now been carried out using a transport phase extractor in which solvent velocities of the order of 0.3 ms^{-1} have been used. Initial experiments using short residence times ($\sim$600 s) have provided encouraging results.

Solids separation. The plant is designed to allow investigation of a number of different solids separation techniques (gravity separation, cyclones, filters) or combinations of techniques. Successful operation has been achieved and extracts have been consistently produced with ash contents significantly less than 0.1% by weight. Work on the plant and other experimental model systems is being directed towards determination of the effect of solids loading on separation efficiency and the development of improved means of filter cake removal.

Process Investigations

The yield of extract obtained at a given set of operating conditions was found to be similar to that obtained during the laboratory scale investigations using the 'mini-bomb' extraction unit. Ultimate analyses for the coal feed, extract and residue for an operation in which a 32.5% daf extract yield was obtained, are given in Table 2, together with ash contents and extract molecular weight and softening point.

Analysis of product gases has shown them to have a consistent composition, the major constituents being methane, carbon monoxide and carbon dioxide. Hydrogen sulphide represented less than 2% by volume of the gas yield which has an average value of 1.5% daf

4

coal feed. Material recoveries in initial plant operations were poor but improvements in procedures have resulted in mass balance closures of the order of ± 7%. Element balances are obviously greatly influenced by the overall mass balance but the oxygen and sulphur balances showed more significant losses of the order of 30%. Specific investigations are now being carried out to improve these balances.

OTHER RELATED STUDIES

As part of the development of the overall processing scheme, a number of other studies are being performed with the aim of providing information regarding other unit operations in the extraction process, evaluation and further processing of extract and solid residues, and assessment of pilot and commercial scale plants.

Extraction Stage

For some time the laboratory-scale experimental programme has been aimed at evaluating multicomponent solvents which more closely resemble solvent species which could be obtained from the second stage hydrocracker products. Considerable success has been achieved and extract yields of the order of 50% by weight of daf coal have been achieved. Recent plant operations have successfully utilised such solvents.

One vital unit operation for which the 5 kgh^{-1} plant does not provide experimental data is that of removal of solid residue from the high pressure system. This is recognised as being the most critical item which has to be developed and an experimental investigation is being carried out in a separate rig. After surveying and assessing a number of systems this rig has been designed to evaluate a scheme in which the residue is reslurried in a fluid and depressurised across an orifice.

Characterisation and Further Processing of Products

Extensive analytical studies (7,8) have been carried out with regard to the structural analysis of the extracts. Analysis of the aromatic fraction, totalling over 95% of the extract, indicated that it possessed a simple open chain structure which could theoretically be broken down to simple, high value aromatics by conventional hydrocracking techniques. Furthermore the nature of the bridging structures suggest that less hydrogen would be consumed than that required to crack condensed aromatic structures into simple units. This should lead to important reductions in the cost of further processing of the extract.

Extract hydrocracking has been performed in a number of batch and continuous processing systems. Most recently experiments have been carried out using a 0.08 dm^3 capacity system operating in a 'trickle' bed mode. Table 3 gives the analyses of the extract feed and product oil from a typical single stage treatment. Information has also been gained on the further treatment of the heavy, +623 K, liquid product, which suggests that this can be further cracked to light liquids. The distillate products have been analysed and appear to be a valuable source of gasoline components and chemical feedstocks. These results have generally been substantiated in a 2 dm^3 capacity hydrocracking unit which is used for studies on coal extracts from both the SGE and LSE processes.

The residual char from the extraction stage maintains the discreet particulates nature of the coal feed (see Table 1) and analysis of the char in terms of proximate analysis, calorific value, and pore structure showed that it could prove a valuable fuel for the production of process heat, power and hydrogen required within an integrated processing scheme.

Thermobalance studies have shown that gas extraction residues are at least as reactive as the parent coal. Further gasification studies in steam, air and carbon dioxide have been carried out in 0.04 m and 0.15 m diameter fluidised bed units which confirm the potential of the residues as gasification feedstocks. The suitability of the residue for combustion has been examined by carrying out ignition and burn-out tests in further tests in a 0.15 m diameter fluidised bed combustor.

Pilot Plant Studies

In response to a proposal from the National Coal Board the UK Department of Energy has provided part sponsorship for a programme of work planned to culminate in the construction and operation of a 25 tonne per day (coal feed) SGE pilot plant. A design study was initiated in February 1979 in order to develop a detailed plant specification and it is intended to proceed to the detailed design and construction stage in 1980. The overall programme calls for the commissioning of the plant, which includes extraction and hydrocracking operations, by the end of 1982.

General Studies

A number of studies have been carried out both in-house and in conjunction with consultants in order to develop and evaluate conceptual commercial-scale processing schemes. A variety of process options, in terms of processing routes and product slates, have been evaluated and comparisons have been made with other liquefaction processes. The majority of these options are based on the principle of generating power, process heat and hydrogen from the residual char, as depicted in Figure 4. Any char excess to requirements in these schemes is converted to synthesis gas.

A process model has been developed which makes it possible to estimate the extract yield required to provide just sufficient residue to satisfy process requirements. This extract yield is estimated to be 47.5% based on daf coal, which is apparently within the process capabilities, and results in the overall process having a thermal efficiency of 67.9%. The overall product distributions per 100 tonnes of raw coal input is given in Table 4.

ACKNOWLEDGEMENTS

Much of the work described here, in particular laboratory and plant scale work on the extraction and hydrocracking stages, was carried out under contracts financed by the European Coal and Steel Community. The paper is published by permission of the National Coal Board, but the views expressed are those of the author.

LITERATURE CITED

1. Davies, G.O., Wyss, W., Gavin, D.G., "Making Substitute Distillates from Coal for the Petroleum and Chemical Industries". ECE Symposium on Gasification and Liquefaction of Coal, Katowice, Poland, April 1979.

2. Paul, P., Wise, W.S., "The Principles of Gas Extraction" Mills and Boon Ltd., London, 1971.

3. Diepen, G.A.M., Scheffer, F.E.C., J.Am. Chem. Soc. 1948, 70, 4085.

4. Whitehead, J.C., Ph.D. Thesis, University of Surrey, 1974.

5. Whitehead, J.C., Williams, D.F., J.Inst. Fuel, 1975, 182.

6. Whitehead, J.C., Crowther, M.E., Proceedings of the Second Engineering Foundation Conference on Fluidisation, Cambridge, 1978, 65-70.

7. Bartle, K.D., et al, Fuel, 1975, 54, 226.

8. Bartle, K.D., et al, Fuel, 1979, 58, 413.

WHITEHEAD, J.C.

Table 1 - Size and ash analysis of coal feed and residue

	Coal feed	Residue R*	Residue V*	Total residue (calculated)
Size analysis, % by weight				
+ 500 μm	0	0.4	0.5	0.5
-500 + 355 "		4.2	0.0	0.9
-355 + 212 "	2.1	17.4	0.9	4.3
-212 + 125 "	10.2	29.6	8.7	13.0
-125 + 75 "	23.4	28.0	20.2	21.8
- 75 "	64.3	20.4	69.7	59.5
Ash %, dry basis	13.7	47.3	14.0	21.9
Parts, by weight, dry basis	100	14	55	-

Temperature 723 K
Pressure 20 MPa
Solvent velocity 4.5 x 10⁻³ ms⁻¹
*R denotes collected from reactor, V denotes collected from main receiver.

Table 4 - Commercial plant product distributions
(Based on conceptual 10,000 tonnes/day plant)

	Parts by weight	Energy content (MJ)
Input		
Raw coal	100	100
Output		
SNG	4.0	7.4
LPG	1.2	2.4
C₅ - 473 K liquids	16.9	28.9
473 - 523 K "	5.2	8.9
523 - 623 K "	11.8	20.3
	39.1	
Thermal efficiency, %		67.9

Table 2 - Extract and residue analyses

	Coal feed	Extract	Residue
Ultimate analyses			
C % daf	80.9	82.7	84.4
H "	4.9	6.1	3.7
O "	11.4	8.5	8.6
N "	1.4	1.3	1.6
S % db	1.5	1.0	1.2
Cl % ar	0.3	0.1	0.4
Ash % db	11.9	0.03	18.0
Volatile matter % daf	40.6	-	21.5
Molecular weight		490	
Softening pt, K		90	

Temperature 723 K
Pressure 20 MPa

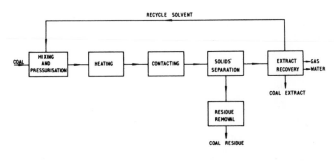

Figure 1. Unit operations in the S.G.E. process.

Table 3 - Analyses of hydrotreated extract

	Extract feed	Product oil
Ultimate analysis		
C % ad	83.3	88.6
H "	6.1	10.7
O "	8.9	0.4
N "	1.3	0.1
S "	0.9	0.1
Removal of oxygen %		95
Removal of nitrogen %		90
Removal of sulphur %		89
Distillation		
-443 K % oil		12
-523 K "		24
-623 K "		55

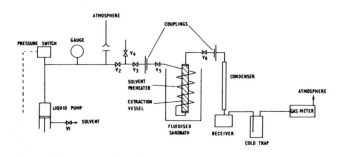

Figure 2. Flowsheet of semicontinuous high-pressure gas extraction rig.

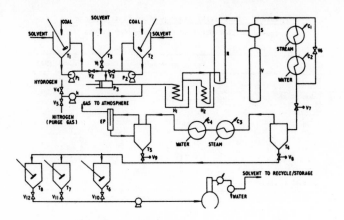

Figure 3. Simplified flowsheet of continuous gas extraction unit (5 kg/h⁻¹).

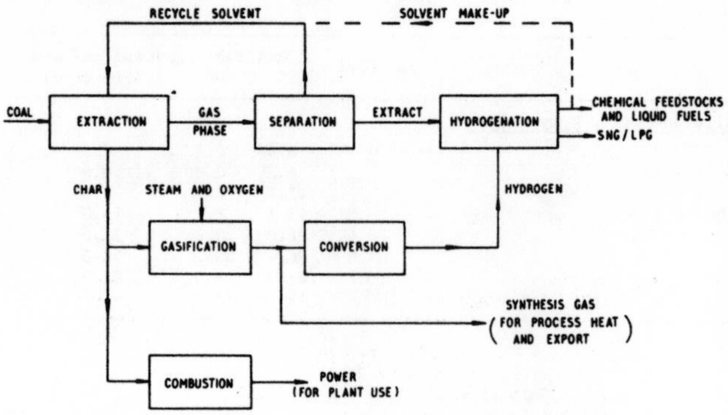

Figure 4. Basic scheme for evaluation of conceptual S.G.E. process.

Transient Model of a Moving-Bed Coal Gasifier

Major factors affecting transient performance are the amount of thermal energy stored in the coal bed, and the rate of drying and devolatilization of the raw coal near the top of the gasifier.

K.J. Daniel
General Electric Co.
Schenectady, N.Y.

A transient model for air- and oxygen-blown, moving-bed coal gasifiers has been developed, and the use of the model shows good agreement with available steady-state data. It was found that the major factors affecting transient performance are the amount of thermal energy stored in the coal bed, and the rate of drying and devolatilization of the raw coal near the top of the gasifier. It was also found that the gas-heating value rapidly increases as the rate of blast input decreases.

The scope of this study was limited to obtaining transient response of the output gas, when changes are made in input blast conditions. Since the model is not intended to produce detailed information of processes occurring within the gasifier, many of the complicated processes are approximated.

Two major approximations of this model are that extremely short-duration transient effects are approximated as instantaneous, and long-duration effects are approximated as being constant. Transients of an intermediate-time scale are modeled in detail. For this reason, the model provides information only for transients on a time scale from one-half to several minutes. Examples of transients which are assumed to occur immediately are: (1) changes in the combustion zone output conditions; and (2) propagation of gases through the reactor. Examples of transients which occur slowly and are assumed to be constant are: (1) the location of the combustion zone; and (2) the height of the coal bed.

In the combustion zone, oxygen is consumed and heat is liberated. In most gasifiers, this zone is small and will respond rapidly to changes in blast conditions. In order to eliminate short-time constants associated with rapid transients and, therefore, make the problem tractable, it is assumed that the combustion zone is at steady state and responds immediately to changes in the blast. In reality, the response of the combustion zone will require approximately 30 seconds.

The steady-state combustion zone model is based on the assumption that no hydrogen is produced in the combustion zone, and the following relations; energy balance, species balances, and pseudoequilibrium between CO and CO_2. The pseudoequilibrium relation was determined by fitting limited data (1) for the peak temperature in a moving-bed coal gasifier.

The gasification zone is modeled using species balance differential equations and a differential energy balance equation. The species equation is solved using only the heterogeneous reactions accounting for species diffusion. The reaction rate constants were those used by Yoon et al. (2). The solution is obtained by using a second-order Runge-Kutta method, integrating from the bottom to the top of the reactor. The initial conditions are determined by the combustion zone model. After each finite step in the solution, water-gas shift reaction equilibrium is imposed.

After integrating the species equation over the axial coordinate, the energy balance equation is used to evaluate the time derivative of temperature for each axial step. A Gear implicit integration method was used to integrate overtime.

In the devolatilization zone at the top of the gasifier, coal moisture and volatile products are assumed to be given off in proportion to the instantaneous coal consumption, lagged by a time constant.

The transient results for a 40% step decrease in blast flow rate are shown in Figure 1 for the conditions corresponding to a full-scale gasifier with a blast steam/air mass ratio of 0.2. This figure shows the raw gas (gas exiting the gasifier) heating value and the heating value of the gas leaving the gasification zone. Note that the transient imposed is rather severe. Nevertheless, this figure illustrates the processes occurring following a transient.

The figure shows the raw gas heating value predictions for time constants of both 90 s and 0 s. The value of the time constant affects only the rate of decay of the effect of the drying and devolatilization zone. It does not affect the size of the effect. The size is determined by the amount of volatile matter in the coal and the magnitude of the change in gasification zone carbon consumption. The size of the transient effect caused by the devolatilization zone is the difference between the results for 0 s and 90 s. This effect is approximately the same size as that cause by the gasification zone.

The immediate increase in heating value of the gas leaving the gasification zone is cause by the relatively high initial temperature of the bed compared to its eventual steady-state value. The initial high bed temperature results in a large rate of reaction for the carbon steam reaction. As a consequence, a larger percentage of the steam reacts forming H_2 and CO, thus increasing the percentage of these constituents in the raw gas.

0149-3701/81/3860 $02.00 © 1981 AIChE

As the bed cools to its final temperature profile, less of the steam reacts and consequently the heating value of the gas drops. Surprisingly, the heating value of the gas does not return to its original value. There are two reasons for this. The first is that the methane production in the gasification zone does not decrease as much as the other constituents of the gas. The second is that slightly more water is reacted at the lower blast flux, because of the longer residence time within the reactor.

Figure 2 shows the change in the raw gas and gasification zone flow rates of the individual species. The molar flow rate of CO, CO_2, and H_2 exiting the gasification zone are almost identical to the flow rate of these species in the raw gas, because devolatilization does not produce large quantities of these species. Consequently, the gasification zone molar flow rates of these species are not shown on this figure.

Some interesting effects can be seen. The gasification zone shows a proportionately larger drop for water than carbon monoxide or hydrogen. The larger drop is due to the previously explained interaction between the bed temperature and the rate of the carbon-stream reaction. The amount of methane leaving the gasification zone is also affected by the transient. Initially, the rate of formation decreases slightly due to a decrease in the hydrogen flow rate. Following this, there is a long-term decrease in the amount of methane produced caused by the propagation of the thermal wave up the bed. Since methane is produced only in the relatively cold upper regions of the gasification zone, methane transients are not seen as quickly as those of other species.

Literature cited

1. Hebden, D., "High Pressure Gasification Under Slagging Conditions," Seventh Synth. Pipeline Gas Symp., Chicago (1975).
2. Yoon, H., J. Wei, and M.M. Denn, "A model for Moving – Bed Coal Gasification Reactions," *AIChE J.*, **24**, 5, 885 (1978).

DANIEL, K.J.

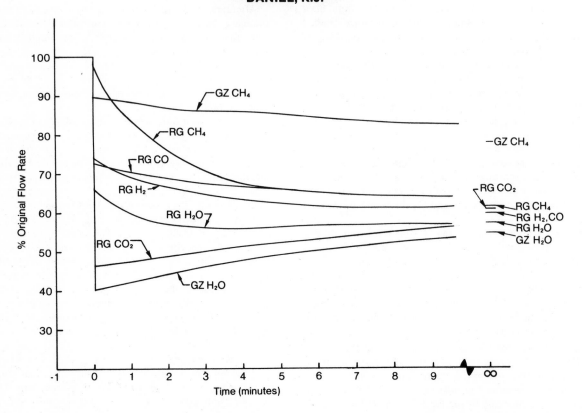

Figure 1. Heating value of raw gas (RG) and gas exiting the gasification zone (GZ) following a 40% step decrease in blast.

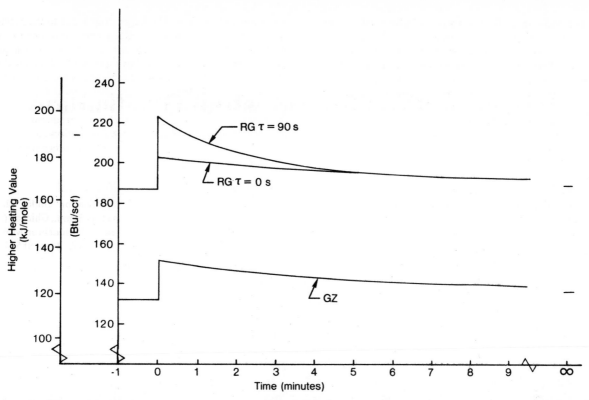

Figure 2. Change in the flow rate of each species following a 40% step decrease in blast (RG, raw gas; GZ, gasification zone exit).

Catalyzing Coal for Industrial Gasification

Battelle Treated Coal offers an economically attractive option for industrial coal gasification currently utilizing fuel oil or natural gas.

B.C. Kim, H.F. Feldmann, L.M. Curran, and C.J. Robb
Battelle, Columbus Laboratories
Columbus, Oh.

Battelle has developed a process for treating coal with calcium compounds (such as CaO), whereby the coal is made nonagglomerating, has its gasification reactivity greatly enhanced, and, by incorporating sufficient calcium in the coal, sulfur is captured during gasification.

Because of these advantages, Battelle Treated Coal (BTC) could be a particularly desirable feedstock for fixed-bed gasification systems currently commercially available. These gasification systems, ordinarily used for industrial applications, must pay a heavy cost penalty for sulfur removal and wastewater treatment and have difficulty with agglomerating coals. Use of BTC could make possible a more rapid conversion of industrial energy systems from natural gas or oil to coal because of fewer institutional, regulatory, and siting problems that delay the construction of large gasification plants.

This work was undertaken for the Department of Energy as a bench-scale screening study to evaluate the technical feasibility of producing low-sulfur gaseous products from gasification of various BTC samples and to assess the economical potential of BTC utilization in small commercial gasifiers.

Specific objectives of the study were to demonstrate that coal treated by BTC could:

- Capture sufficient sulfur in the ash to eliminate the need for gas desulfurization.
- Produce an ash from which the sulfur would not be leached.
- Eliminate the potential for agglomeration.
- Enhance the gasification reactivity of the coal.

Experimental work demonstrates that these objectives could be achieved. Economic projections are made comparing the use of BTC in fixed-bed gasifiers with other means of coal utilization for industrial steam generation. The coal used in all of the experimental work is Illinois No. 6 from Christian County.

BTC is prepared by impregnating the coal with CaO utilizing an aqueous slurry. (This is optional.) Figure 1 shows the basic elements of the treatment process, where the treatment system also doubles as a slurry feeder at a pressurized gasifier. For "conventional" gasification systems (1), the treatments are primarily intended to eliminate the agglomerating tendency of the

coal and to increase its reactivity. For this purpose, the concentration of CaO in the coal is typically from 2 to 5 wt.%. However, to capture sulfur, considerably more CaO is required. Therefore, simple dry mixing with coal of CaO as well as Ca(OH)₂ and

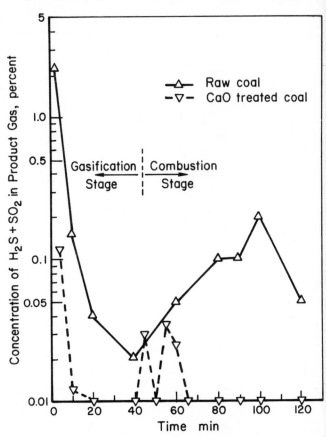

Figure 1. Hydrogen sulfide and sulfur dioxide generation in fixed-bed gasification of Illinois #6 coal.

$CaCO_3$ were compared with slurry treatment utilizing the above calcium compounds.

Separate bench-scale gasification tests were conducted to determine reactivity of the coal, its agglomerating behavior and most importantly *in-situ* sulfur capture. The sulfur capture of BTC in bench tests simulating a fixed-bed, air-blown gasifier is compared with raw coal in Figure 1.

Major experimental conclusions of this study are:

- With dry mixing of coal with the treatment agent only, $Ca(OH)_2$ was effective in eliminating agglomeration. However, with slurry treatment all compounds eliminated coal agglomeration.
- Removal of 80 to 85% of the sulfur in the coal (the original sulfur level was approximately 5%, in the ash could be achieved at a Ca/S mole ratio of approximately 3.
- The ash in the sulfur from a fixed-bed gasifier was found nonleachable and therefore its disposal should not create environmental problems.
- Steam concentration is an important variable influencing sulfur retention by the ash with reductions in steam concentration resulting in increased sulfur retention.
- Gasification and apparently oxidation reactivity of the coal was greatly increased by the presence of calcium compounds.

Because the experimental goals were met, a conceptual economic evaluation was made of the cost to retrofit an oil-fired industrial installation requiring an overall average of 27,240 kg/h (60,000 lb/h) of steam of the following four options:

(1) Coal-fired boilers with fluid-gas desulfurization (FGD)
(2) Gasification of untreated coal with the hot producer gas utilized in existing boilers followed by FGD.

(3) Gasification of untreated coal with sulfur removal from the producer gas.
(4) Gasification of BTC prepared and briquetted at the mine.

The results of this economic evaluation are summarized below.

Options	Coal-Fired Boilers with FGD	Gasifiers with FGD	Gasifiers with H_2S Removal	Gasifiers with BTC
Total Capital Investment, $M	9.1	9.8	10.4	7.7
Annual Operating Costs, $M/yr	2.8	3.1	3.1	3.0
Steam Cost $/454kg (1000 lb)	10.7	11.6	11.9	10.3

Thus, BTC offers an economically attractive option for the implementation of industrial coal gasification currently utilizing fuel oil or natural gas. In addition to the economic advantages, the utilization of BTC would be much simpler than the above options, because it eliminates the need for complex processes for gas phase sulfur removal.

Literature cited

1. Feldmann, H.F., S.P. Chauhan, P.S. Choi, and H.N. Conkle, "Gasification of CaO Catalyzed Coal," 14th Intersociety Energy Conversion Engineering Conference, Boston, MA (August 5-10, 1979).

CURRAN, L.M.

FELDMANN, H.F.

ROBB, C.J.

KIM, B.C.

Operation of the Bi-Gas
Coal Gasification Pilot Plant

This plant was designed to convert all types of coal into high-Btu pipeline quality gas. Detailed procedures for start-up, operation, and shutdown of the gasifier and peripheral units are presented along with a generalized future program.

J. Glenn
Stearns-Roger Engineering Corp.
Homer City, Pa.

The BI-GAS Pilot Plant is a grass-roots facility designed to convert up to 4535 kg/hr (5 tons/hr) of coal into high BTU pipeline-quality gas. A principal feature of the plant is a two-stage, entrained flow, high pressure, oxygen-blown, and ash slagging gasifier unit. This unit was scaled up from a bench scale facility designed and tested by Bituminous Coal Research, Inc. (BCR). The pilot plant, which has been operated since December 1976, is currently solely funded by the U.S. Department of Energy.

Significant progress has been made over the past year in the BI-GAS plant operation characterized by smoother start-ups relative to previous years, longer test runs, and planned shutdowns when required process/operating data had been collected. The past operating problems of taphole pluggage, performing consistent and safe burner ignition at pressure, slag removal, and solids sampling for material balance purposes have essentially been eliminated. Improvements have also been noted in char burner design and in the various supporting units' operation.

The purpose of this article is to present (1) a description of the current status of each plant unit's operation, and

(2) a description of the gasifier start-up and shutdown procedures. Refer to igure 1 for a general schematic and flow scheme of the BI-GAS plant.

COAL PREPARATION (Figure 2)

Rosebud coal is fed into a rod mill through a screw conveyor where it is slurried with water to approximately 20 weight percent and ground with twenty 0.102m (4 inch) diameter rods. The rod mill effluent is then pumped to a Rotex vibrating screen unit with a 100 mesh screen where the slurry fines concentration (minus 100 mesh) is reduced from approximately 30% to 15%. The top-of-screen product is then reslurried to a 35% to 40% solids concentration and stored in a large tank where it can be pumped to the spray drying section. The "waste" (minus 100 mesh slurry) is drained to the waste treatment pond. Engineering work is being done to provide the capability to concentrate this stream into a cake for more convenient disposal. Typical size analyses of the rod mill discharge and product slurry are shown on Table 1.

Recent addition of a screw conveyor has essentially eliminated the pluggage problems occurring around the bucket elevator which had previously been used to transport the coal to the rod mill. Tests have shown that the conveyor is capable of transferring the coal at more than double the current grinding rate of approximately 4535 kg/hr (5 ton/hr). Although the newly-installed coal screen does not remove all the minus 100 mesh material, it does "coarsen" the grind to the extent that the downstream cyclones are operating more efficiently. Operating tests on the present screen have not further improved the fines removal at the present desired slurry feed rate. It appears that, as suggested by the vendor, a duplicate screen in parrallel with the present unit would be required to further reduce the fines concentration. However, since the operation of the downstream units have been adequately improved with one screen, there is no further incentive to reduce the fines concentration.

SPRAY DRYING SECTION (Figure 3)

Slurry from coal preparation is pressurized to slightly greater than gasifier pressure at 5275 kPa (765 psia) with a triplex plunger pump. It is then heated to approximately 206C to 234C (400F to 450F) with a steam preheater; then contacted with a hot gas stream, where the water is vaporized and the combined stream reaches a net temperature of 290C (550F). The vaporized water and gas are then separated from the dried coal in a cyclone. The gas stream is then water washed (where the bulk of the vaporized water is condensed), compressed, and heated for return to the drying step.

The flash dryer itself has performed well during the entire operational period of the plant. Since the coal screen has been used, the cyclone efficiency has increased from approximately 75% to 95%. The resulting net reduction of solids in the gas washer discharge liquid stream (from 10% to 1%) has improved operation by reducing pluggage in process lines and instrument taps. Table 1 shows a typical size analysis of the waste stream leaving the spray dryer gas washer.

Line plugs have occurred more frequently in the coal slurry lines up and downstream of the slurry pump after system shutdowns because of the coarser size distribution of the coal. These plug formations have necessitated immediate flushing of these lines and dilution of slurry in vessels to prevent the solids from settling out once the system is shut down. In general, the spray drying system has performed adequately with a minimum of operational problems.

GASIFIER SECTION (Figure 4)

Figure 5 presents a sketch of the BI-GAS gasifier. The gasifier is a two-stage entrained flow, high pressure [5275 kPa (765 psia) present operating pressure; 10,446 kPa (1515 psia) maximum], slagging, and oxygen blown unit to convert coal into a methane rich gas. Unreacted char from the coal gasification section (Stage II) is quenched to 430C (800F), separated from the product gas with cyclones, then recycled to the lower section (Stage I) for reaction with oxygen and steam to produce heat and hydrogen for the gasification step. Slag from Stage I and synthesis gas from the char cyclones are the only products leaving the gasifier area.

Stable and proper operation of the gasifier has been the primary emphasis of the BI-GAS Pilot Plant operation since construction completion in mid-1976.

A present operating description and status summary of major sections of the gasifier are presented below:

Coal Feed System (Figure 6)

Stage II (upper section of the gasifier) is fed with dry pulverized coal through two coal eductors and nozzles (each eductor/nozzle being 180° from the other) using recycle gas as a motivating fluid. Coal is fed to the suction side of each eductor, through a 0.102m (4 inch) diameter vertical leg connected to the coal feed vessel which normally contains a level of coal. A Fisher Vee-ball control valve, operated from the control room, regulates the flow of solids through the eductor. The coal vessel is maintained at a 35 to 69 kPa (5 to 10 psi) higher pressure than the gasifier.

The eductor merely "pumps" the coal through a nozzle tilted upward 30° from horizontal into the gasifier. An outer annulus on the nozzle feeds steam for reaction and temperature control purposes. Since the coal/steam, coal/hydrogen reaction is endothermic, steam flow to Stage II is reduced (via a temperature controller) as coal flow is increased. The upper section of Stage II is maintained in this manner at 906C (1650F); the lower section of Stage II generally operates 38C to 94C (100F to 200F) hotter.

Coal feed is detected through a variety of instrumentation. Three differential pressure instruments (across the coal leg, across the eductor, and across the coal nozzle) serve to detect the presence of solids feed or no solids feed. A coal leg thermocouple, located in the line immediately above the eductor, senses the inside line temperature. The coal vessel nuclear level detector will show flow and change of feed rate that will result in a change in coal vessel level. Efforts are currently being made (although not complete) to relate level and change of level to a mass of coal. A density indicator located on the coal leg above the eductor may show a filled coal leg in the event of a plug or flow down the line.

Typical responses of instruments to initiation of coal feed are as follows: reduction of the coal leg and eductor differential pressure (dP); increase of the nozzle dP; increase of the coal leg temperature as a result of "hot" coal flowing through the leg (several purge gas taps connect to the coal leg for expansion joint purging and dP instrument purges which cool the line when coal is not flowing); approximately 10% reduction in the purge gas flow to the coal eductors caused by the eductor loading being increased; and reduction in the amount of steam being fed Stage II. The above instrumentation is used primarily as a flow/no flow indication. The mass coal rate to the gasifier is calculated by making a material balance calculation around the spray drying section of the plant.

Coal feed rate to the gasifier has been limited to a 907 to 1814 kg/hr (2000 to 4000 lb/hr) rate because of restrictions and plugs that have formed in the coal leg, eductor, and nozzle. Since the restrictions have occurred in a variety of locations, it has been difficult to assess the specific cause(s) of the problem. Potential problem areas that have been considered include water condensation that may be occurring in the coal leg; the relatively large control valve may be feeding too much coal to the downstream eductor or nozzle at minimum or small valve opening; capacity limitations of the eductor; and inadequate coal/purge gas velocity in the coal nozzle to prevent settling out of the coal, thus causing restrictions. Calculations, however, show this velocity to be approximately 7.62 m/sec (25 ft/sec), which is felt to be adequate for pneumatic conveying of the coal.

Two major changes were made in January 1980 following Test G-10A to eliminate the coal system feed limitation. A 0.038m

(1.5 inch) diameter vertical coal leg was installed on one of the two coal feed lines to replace the original 0.102m (4 inch) line.

The 0.076 (3 inch) Vee-ball control valve was replaced with an 0.025m (1 inch) Vee-ball valve. Refer to Figure 7 for a drawing of the new leg. It is felt that this line will improve feed for two reasons: The smaller coal line will not require expansion joints and will thus require a minimum of purge gas flow to the leg (only that amount required to purge instrument taps). Less purge gas will result in a warmer leg temperature less susceptible to condensation. Previous test data indicated that the leg temperature was running very close to the dew point of the spray dryer gas entrained in the coal. A smaller control valve will provide more sensitive flow control and will not "slug" the eductor with large quantities of coal. The second change made was the addition of a purge gas heat exchanger. This unit, through heat exchange with 6895 kPa (1000 psia) steam, will heat all of the purge gas feeding the gasifier to approximately 150C (300F). This hotter gas should further eliminate potential for condensation in the coal and char feed systems.

Char Feed System (Figure 8)

Stage I is fed with char through three nozzles (120° apart) utilizing an eductor and steam as the motivating fluid. The arrangement is physically similar to the coal feed system. Each char burner is in a horizontal position and has several annuluses as described below (see Figure 9 for char burner assembly drawing); center tube for char, motive steam, and purge gas from instrument taps in the leg; first annulus for supplemental fuel gas and additional reaction steam (sufficient fuel gas is currently fed to the gasifier to stoichiometrically consume all the oxygen fed Stage I); second annulus for oxygen; and the third annulus for cooling water to protect the char burner tip.

The char feed problems associated with plugs in the burner and leg have been eliminated as a result of several modifications completed in late 1979: replacement of the original 0.102m (4 inch) diameter line and 0.076m (3 inch) Vee-ball control valve with a 0.025m (1 inch) diameter line and 1" Vee-ball valve (This new line, presently installed on two of the three char feed systems, has reduced the purge gas requirements due to the elimination of expansion joints and has improved the solids feed control.); the installation of a fuel gas heater designed to heat the fuel gas to 246C (475F) and thus eliminate the potential for condensation of steam in the char burner; the installation of a smaller center tube in the char burner resulting in the configuration described above (This new center tube separates the fuel gas from the char/motive steam such that if a plug forms in the center tube, fuel gas flow will not be stopped.).

Instrumentation for detection of char feed is somewhat more extensive than that on the original 0.102m (4 inch) coal leg, described earlier in the report. The major instrumentation is presented below. Differential pressure (dP) taps measure and record the following dP's; top of char vessel to the top of the Vee-ball valve, top of the Vee-ball valve to the bottom of the valve, bottom of the Vee-ball valve to the eductor suction, eductor suction to the char burner inlet, and char burner inlet to the gasifier. Tests continue in order to understand the dP instrument response to flow/no flow or magnitude of char feed. The Argonne acoustic flow monitor is located downstream of the eductor and has been shown to give a char flow/no flow indication. Thermocouples are located in the top and bottom of the vertical section of the char leg. As in the coal system, the thermocouples respond to the "warming" of the leg with improved feed. The char vessel nuclear level indicator and leg density indicator provide similar information to those instruments located in the coal leg.

Char feed is manually controlled through adjustment of the Vee-ball valve to maintain a stable level of char in the char vessel. The oxygen flow (on automatic flow control) is adjusted as necessary to control Stage I temperature at 1494C to 1550C (2700F to 2800F). Motive and process steam and fuel gas flows are held constant.

Slag Removal System (Figure 4)

Slag flows from the sides of Stage I through a 0.051m (2 inch) diameter taphole (generally in one to two streams) into a water bath maintained at approximately 66C (150F). Water in the bottom of the gasifier is continuously pumped through the one lockhopper in service (connected to the gasifier via a 0.3m line), through a heat exchanger for cooling, then back into the gasifier. A mechanical agitator (consisting of four blades at the bottom of the gasifier) alternately rotates 90° clockwise and counterclockwise to break up any slag strands that form in the water and aids in the transfer of slag to the lockhoppers. The slag particles then flow into the lockhopper where they are pumped to a collection vessel for manual weighing. No serious problems have been recently encountered in this system's operation. Slag of proper consistency is reliably removed from the gasifier with the taphole remaining open throughout a test.

The internal weigh cells installed in the slag lockhoppers could not mechanically withstand the pressure cycles of the lockhoppers and are no longer in service. A stalactite breaker, which is merely a water-cooled mechanical arm, rotates underneath the slag taphole to break off any stalactites that may form below the taphole. A television monitor which views the slag taphole from the bottom section of the gasifier, has shown the breaker to effectively "knock-off" stalactites before they reach a large size. The quality of the television picture has consistently been good during the past year of operation.

GAS TREATING SYSTEM (Figure 10)

Synthetic product gas leaves the char cyclone vessel where the gas is contacted with water in the product gas washer and cooled to 177C (350F), cooled to approximately 27C (80F) with water and naphthalene being knocked out, fed to the H_2S absorber where the stream is countercurrently contacted with Selexol for H_2S removal, and fed to the thermal oxidizer for disposal.

The methanation and CO_2 absorption sections are presently available for operation but have not been run to date due to emphasis of achieving stable gasifier operation.

The gas washer system has operated quite satisfactorily as a result of minimizing the amount of char fines being fed to this unit. Table 1 shows a typical size analysis of the waste liquid stream leaving the product gas washer.

Physical operation of the gas washer with (1) the recycle stream removed from the vessel bottom, pumped to the top of the tower, and cooled; and (2) the bottoms slurry stream being removed on level control has been quite stable. Pluggage problems in the gas washer streams have been minimized since the coal grind was altered. Ammonia concentration of the bottoms stream has been effectively reduced through the operation of the downstream low pressure vent gas washer as a steam stripper. Steam is added in the lower section of the low pressure tower for proper stripping while caustic is added to raise the pH of the bottoms liquid to eleven. Typical ammonia concentration of the major slurry stream entering the stripper is 1500 ppm while that in the stream leaving the stripper is 10 to 15 ppm.

The synthetic product gas is cooled to approximately 27C (80F) and sent through a vessel where the water particles are disentrained. An aerial cooler in series with two shell and tube heat exchangers (in parallel with each other) serves to cool the gas. Each of the shell and tube units, cooled with plant cooling water, condense the naphthalene material in the gas. An excessive pressure drop in each unit (138 to 207 kPa or 20 to 30 psi) indicates the need to place the parallel unit in service and steam clean the "dirty" exchanger. Steam is fed to both the shell and tube side of the unit which melts and vaporizes the naphthalene. The naphthalene is then sent to the thermal oxidizer. Alternation of the use of these exchangers provides for removal of the naphthalene formed with a minimum of naphthalene being fed to the Selexol, H_2S removal unit. Recent operating history has shown that the need to cycle exchangers occurs once every 24 to 48 hours of operation.

The H_2S removal system has performed with a minimum of problems in the past year. This H_2S absorption system typically reduces the H_2S concentration of the product gas from 1500 to 2000 ppm to less than 1 ppm. Typical gas loading rate of $7079 m^3/hr$ at 15.6C and 101.4 kPa (250,000 SCFH).

GASIFIER START-UP AND SHUTDOWN PROCEDURE

Normal start-up of the gasifier is simple in concept and requires a minimum of time. Several activities in the plant are initiated before pressure-up: circulation of the spray drying system gas washer and the product gas washer is started; water circulation through the slag lockhoppers is started; gasifier cooling water circulation is started; and nuclear level devices are zeroed.

The entire plant is then pressured at a rate of 1034 to 1379 kPa/hr (150 to 200 psi/hr) with nitrogen feeding the spray drying system and purging the oxygen lines to the gasifier. All utilities feeding the gasifier (such as nitrogen, recycle or purge gas, service water, etc.) are maintained at a constant pressure greater than the gasifier through differential pressure controllers. The spray drying gas circulation is started at approximately 689 kPa (100 psia) along with the spray drying gas heater to heat the gas to 290C (550F). Once up to 2068 kPa (300 psia), the Selexol system circulation is initiated. This minimum pressure is required to allow for proper pressure letdown in the two flash tanks. The gasifier cooling water is heated via a steam coil during pressure-up to a desired temperature of 240C (460F) at 5275 kPa (765 psia).

Once up to 5275 kPa (765 psia) pressure, motive steam flow through the three char eductors into Stage I is started and fuel gas flow to the five gasifier burners (slag tap, slag heating, and three char burners) is initiated. Once the Triethylaluminum (TEA) ignition systems (feeding only the slag tap and heating burners) are put in service, the two burners are lit at pressure through TEA injection. The fuel gas and oxygen flows to these two burners are then brought up to the specified "reducing" conditions. The ignition of each burner, which has been demonstrated to be reliable at pressure, is carried out in several steps (refer to Figure 11 for a drawing of the burner/ignitor). With fuel gas flowing through the burner, TEA and oxygen are simultaneously injected into the burner. Any flame resulting from the TEA, oxygen, and fuel gas reaction is detected by a flame sensor mounted at the rear of the burner. At the end of a prescribed period of time, the TEA injection will stop, the oxygen flow will continue (if the flame sensor detects a flame) or the oxygen flow will stop (if the flame sensor does not detect a flame). The fuel gas flow is continuous. Any loss of flame detection will automatically shut the burner down.

Past problems with TEA line pluggage, proper flame detection, and inconsistant operation of the ignition system have been basically solved. The TEA system has demonstrated its reliability in the past one to two years of operation. Many safety-oriented procedural steps have been incorporated in the ignition system operation to minimize the hazards related to this pyrophoric chemical.

The following steps are then followed to complete start-up of the gasifier. A minimum of steam flow is started into Stage II with the flow put on automatic temperature control. The gasifier overhead quench section water controls system is set up to control the Stage II overhead temperature at 430C (800F). The three char burners are lit (one at a time) by using the slag heating burner as a pilot light. Successful ignition of each burner is detected by an increase in Stage I temperature. Once the char burners are lit, firing is increased over a period of approximately one hour to 1102C (2000F) in Stage I. Immediately after char burner firing, coal slurry feed to the spray drying section is started, to accumulate a level of dried coal in the coal vessel. The Vee-ball valves below the vessel are closed at this point. It typically takes one hour to observe a level increase in the vessel. Once a level is detected, the char burner firing rate is increased (through oxygen flow) to achieve 1494C (2700F) in Stage I. Process steam is then added, resulting in "base" conditions being reached. Base conditions are achieved when the gasifier is at normal operating temperature, with normal gas flows and with no solids feed. Once a normal level is reached in the coal vessel, the char leg Vee-ball valves are closed and coal feed is initiated through both coal legs. The coal feed rate is adjusted to maintain a stable level in the coal vessel. Once steady coal feed has been established for one hour, a minimum char flow is started through one char leg. This is done to maintain an adequate coating of slag on the Stage I cooling coils. After a char level is detected in the char vessel (typically 5 to 10 hours after coal feed initiation), char feed is started through the three char legs to maintain a stable level in the char vessel. At this point, normal operating conditions have been established. Typical time required to meet test conditions after the plant is pressured up is 12 to 18 hours.

Gasifier shutdown is summarized below to the point where the plant is to be depressured. Coal slurry feed to the spray drying section is slowly reduced to zero, resulting in no feed to the coal vessel. Spray dryer heater firing must be reduced correspondingly to account for the reduced heat load. Coal feed to the gasifier is stopped by closing the coal leg Vee-ball valves once the desired level is achieved in the coal vessel. Steam is automatically increased to maintain Stage II temperature at 906C (1650F). Once depressurized and allowed to cool, the coal

is drained from the vessel and weighed. This weight combined with the level reading in the vessel at test end is plotted on the level calibration curve being formulated. Char feed is stopped by slowly closing the control valves. Stage I temperature is maintained at 1494C (2700F) by reducing the oxygen flows to each char burner. A level calibration curve for the char vessel is being formulated in a similar manner to that of the coal vessel. Char burner firing rate is reduced to 50% at which point the burners are shut down. After firing for one additional hour, the slag tap and heating burners are then shut down. Fuel gas and steam flows are stopped. Depressurization procedures are then begun for the plant.

The gasifier section of the plant is instrumented with an emergency shutdown (ESD) capability when important process parameters exceed certain limits. The ESD simply stops coal and char feed, stops oxygen flow, and initiates a nitrogen purge to all five burners. All remaining flows continue into the gasifier for purging and cooling. Once the ESD cause is thoroughly investigated and eliminated, the gasifier can be brought back up to temperature with normal operation being resumed.

GENERAL FUTURE PROGRAM

The immediate emphasis of the BI-GAS program is directed toward operation of the plant in order to allow for evaluation of the process. Steps which will be taken in the next year to achieve this goal include the following: continued evaluation of the new char feed system to confirm stability of char feed and proper char flow rate, if successful, the third char leg will be converted over to the new system; evaluation of the coal feed system with the original design and the new design completed in January and later revised, if necessary, to provide stable and sufficiently high coal flow rates; demonstration of the operability of the plant with the new solids flow systems during a long-duration test; continued work to test commercially available and BI-GAS-developed solids flowmeters, Stage I thermocouples, and other specialized instrumentation; development of a program to eliminate or reduce supplemental methane feed to Stage I; demonstration of the operability of the plant with reduced or zero methane feed during a long-duration test.

GLENN, J.

TABLE 1

Typical Size Analyses of BI-GAS Plant Slurry Streams

U.S. Sieve No.	Rod Mill Effluent	Product Slurry	Waste Stream Leaving Spray Dryer Gas Washing System	Waste Stream Leaving Product Gas Washing System
8	2.9	2.6	-	-
12	5.4	5.0	-	-
20	20.9	19.4	-	-
50	30.4	41.7	5.4	-
100	12.8	17.4	25.9	0.6
200	9.5	6.9	21.5	1.4
325	4.4	1.5	6.9	2.0
-325	13.8	5.5	40.2	96.0
Typical percent solids in slurry	20%	35%	1%	0.25%

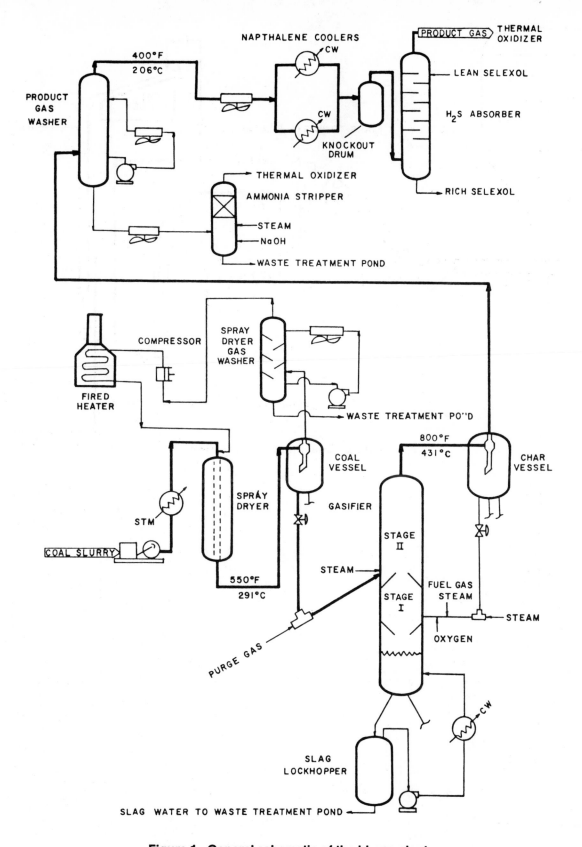

Figure 1. General schematic of the bi-gas plant.

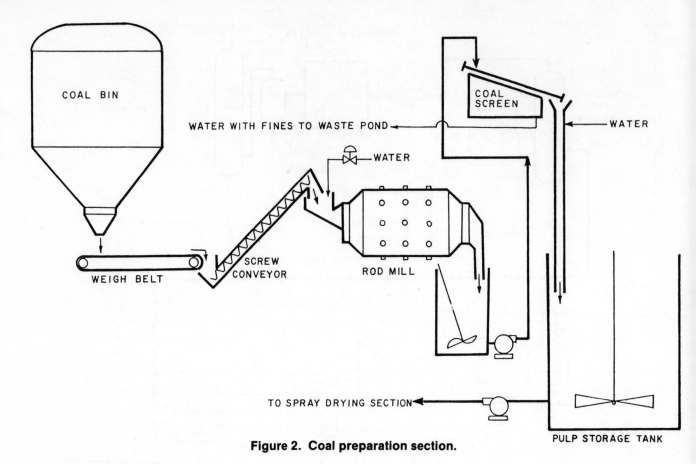

Figure 2. Coal preparation section.

Figure 3. Spray drying section.

Figure 4. Gasifier section.

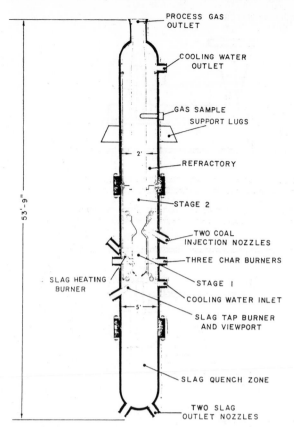

Figure 5. Bi-gas gasifier sketch.

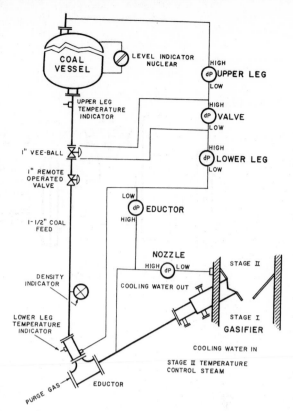

Figure 7. 1.5-in. (0.04-m) coal feed system (SI conversion: m = in. × 0.025).

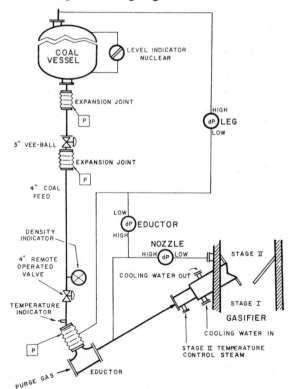

Figure 6. 4-in. (0.1-m) coal feed system (SI conversion: m = in. × 0.025).

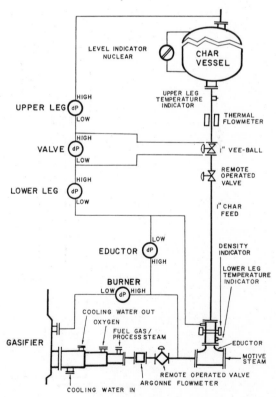

Figure 8. 1-in. (0.025-m) char feed system.

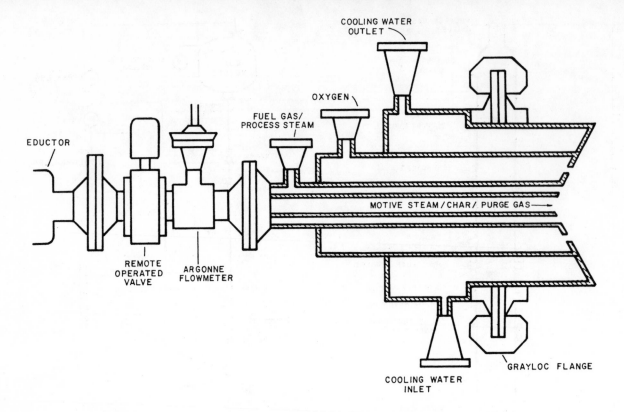

Figure 9. Char burner assembly.

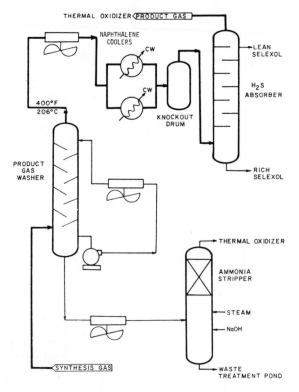

Figure 10. Gas-treating system.

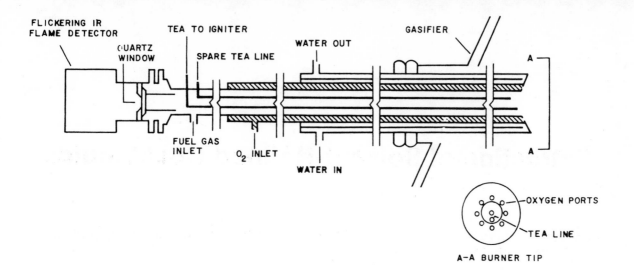

Figure 11. Burner/ignition system.

Extraction of Solvent-Refined Coal Liquids

Extraction of phenolics and asphaltenes appears to be an important step in upgrading coal liquids to refinery charge stock.

E.G. Scheibel, A. Schneider, and E.J. Hollstein
Suntech Group
Marcus Hook, Pa.

The liquefied coal product of the SRC-II process contains large quantities of oxygen and nitrogen compounds which make it unsuitable as a charge stock for existing refineries. Hydrogenation has been considered for reducing the heteroatom content of the coal liquids to acceptable concentrations. Relatively long contact times at severe conditions are required. Hydrogen consumption is also excessive because, under these conditions, the polynuclear aromatic hydrocarbons are converted to paraffinic gases and naphthenes. This not only reduces the potential yield of gasoline but also requires recovery of the hydrogen content of the gases. Efficient hydrogen utilization is the most important economic factor in converting coal liquids to refinery charge stock.

Solvent extraction can provide an alternate approach to upgrading coal liquids by selectively removing the heteroatom compounds from the hydrocarbons. In this study, the distribution coefficients of the heteroatom compounds present in solvent refined coal liquids of different boiling points were determined. The objective was to evaluate the extraction process for maximum removal of these compounds from SRC-II liquids and prepare samples of raffinates and extracts for

further hydrotreatment studies. The results of the hydrogenation work will be reported at a later date.

The recovery of phenolic compounds and other chemicals may also be economically important in the early stages of development of coal liquefaction processes. When coal becomes a significant source of synfuel, the quantities of these chemicals available in the coal liquid will far exceed the demand. Their ultimate use will require conversion of thesse compounds to fuel components.

DISTRIBUTION DATA FOR AQUEOUS METHANOL SOLVENT

In the initial exploratory work on the extraction of SRC-II coal liquid, the middle distillate fraction (199° - 288°C boiling range) was subjected to a fractional liquid extraction with 1.5 volumes of methanol containing 15 volume percent water in an 11-stage operation with feed to the center stage and refluxing about half of the extract product by adding 0.1 volume of water to the final extract in the top stage. The funnel rack and operating techniques used in these studies have been described by Scheibel (1). When steady state conditions were approached,

0149-3701/81/4017 $02.00 © 1981 AIChE

samples of the light and heavy phases in all equilibrium stages were analyzed for phenol and nitrogen content.

The data indicated that the nitrogen compounds in the coal liquid did not distribute as a single compound in the multistage extraction. Empirical breakdown of the nitrogen compounds into fractions with different average distribution coefficients made it possible to develop equilibrium stage calculations which matched the experimental data within the analytical accuracy. The study was repeated using a composite feed sample of the middle distillate, the heavy distillate (288°C+) and the 177°C+ fraction of the naphtha which thereby provided a charge containing all phenolic compounds in the SRC-II liquid. The heavy distillate increased the viscosity of the hydrocarbon phase and the naphtha fraction decreased the density difference so that phase separation at ambient temperature was too slow to be practical in a continuous multistage column. The equilibrium data in the stages at steady state also indicated a variation in the distribution coefficients of the different nitrogen and oxygen compounds in the feed. Black insoluble material which collected at the interfaces in the stages was suspected of contributing to the poor phase separation.

The middle distillate and the naphtha were separated by fractional distillation into ten fractions each. The middle distillate fractions and the highest boiling naphtha fraction were then subjected to a four-stage liquid extraction with feed to the second stage and water added to the first stage to provide reflux to the feed stage. The feed was thus subjected to three countercurrent extraction stages with 1.5 volumes of methanol containing 15 volume percent water. Reflux to the feed stage was obtained using the same 0.1 volume of water addition to the extract. The reflux operation in this case allowed the concentrations of the extracted compounds to build up to steady state conditions in the feed stage at a greater rate per cycle. This was desirable because feed quantities were limited, and the amount available was sufficient for only four feed cycles. Figure 1 shows the schematic diagram of these extraction operations.

The approach to steady state conditions varies with the extraction factor, i.e., the ratio of the distribution coefficient to the solvent ratio, and after three raffinate product cycles in a three-stage system, the

maximum deviation from steady state is about 10%, and after four such cycles, this maximum decreases to about five percent. By refluxing some of the extract to the feed, the quantity of raffinate will approach a larger value than that obtained in a three-stage stripping type extraction. The assumption of steady state conditions for this type of extraction after only three or four raffinate product cycles is, therefore, reasonable. The mathematical analysis of the technique is similar to that previously described for increasing the rate of approach to steady state in a dual solvent fractional liquid extraction (2).

Mathematical relationships for countercurrent liquid extraction processes are normally based upon the assumption of a constant solvent flow ratio in all stages. When this ratio varies, the proper average value must be used in the simplified mathematical relationships to provide the correct results. The fraction of residual solute in the raffinate after n countercurrent stages at steady state is:

$$\frac{x_n}{x_0} = \frac{E - 1}{E^{n+1} - 1}$$

where x_n = concentration in raffinate after n stages

x_0 = initial concentration in feed

E = extraction factor = VK/L

L = quantity of raffinate phase

V = quantity of solvent phase

K = distribution coefficient in terms of concentration in solvent phase/ concentration in raffinate phase

The average distribution coefficient for a measured fraction of a compound remaining in the raffinate after three stages can be evaluated from a variation of x_n/x_0 with E as shown in Figure 2.

This figure also shows the calculated fraction unextracted after three or four raffinate product cycles. However, since the use of reflux to the feed stage will increase the fraction unextracted above the calculated steady state value after more product cycles, the steady state curve in in this figure was used to evaluate the distribution coefficients for all the experimental data.

Variations in liquid phase quantities in the stages contribute the major uncertainties to the evaluation technique. The ratio of the phase quantities varies depending upon the amount of feed extracted in the successive stages. The weight ratio of solvent feed to raffinate product was used as the average value in evaluating the distribution coefficients from the present data. Material balances on the stages showed that this average value was reasonably accurate when there was very little extraction from the feed and also when about 3/4 of the total extraction occurred in the feed stage. Under intermediate conditions, the actual average solvent ratio might be up to 10% lower than the solvent to raffinate ratio in the last stage indicating that the actual average distribution coefficient could be as much as 10% higher than that evaluated from steady state conditions in Figure 2.

Table I summarizes the distillation data and the extraction data for the SRC-II middle distillate. The data for the total naphtha and heavy distillate fractions and a high boiling naphtha cut are also included in this table. Figure 3 (a), (b), (c), and (d) show the variation of the distribution coefficients for total nitrogen, total oxygen, basic nitrogen, and phenolic compounds, respectively, with the boiling points of the different cuts. The curves represent the integrated average variation with boiling point for the different types of compounds. The curves are surprisingly smooth considering that the individual compounds of any particular type in a given boiling range could have different molecular structures and, therefore, significantly different distribution coefficients.

Comparison of the curves in these figures indicates that the phenolic compounds are extracted most readily and basic nitrogen compounds are the most difficult to extract. The apparent inconsistency in the phenol data in Figure 3 (d) at the high boiling range is due to accuracy limitations of the analytical technique. Phenol was determined by the colorimetric method using 4 amino antipyrene, and in order to eliminate interference of other compounds in the coal liquid, the samples were dissolved in octane and extracted with four volumes of water. This procedure would not extract extremely high molecular weight phenolic compounds and fractionation produced by the analytical technique could account for

the steep slope of the distribution coefficient curve in Figure 4 (d).

The effect of the distribution coefficient variation on the overall extraction of the different heteroatom compounds from coal liquid can be more readily recognized when the variation is considered as a function of the total amount of the given types of compound in the middle distillate. These curves are shown in Figure 4 (a),(b),(c),& (d) with the cumulative percent of each compound in the abscissae in terms of increasing boiling points. The applicability of this variable is based upon the fact that the distribution coefficient shows a consistent variation with boiling point as indicated by Figures 3 (a) - (d). Without this relationship, these curves would show a random variation and the fractions would have to be realigned on the basis of a consistent variation in distribution coefficients for the different types of heteroatom compounds.

Figure 4 (a) also shows an empirical breakdown of the nitrogen compounds into three fractions of different distribution coefficients which was derived from concentrations in the equilibrium phases of a countercurrent extraction system consisting of six stripping stages and four fractionating stages operated at a solvent to feed ratio of 1.5 and a reflux ratio of about 0.5. Reflux was obtained by adding water to the extract product stream equal to 0.1 volume of the feed in a fifth stage above the feed. The empirical variation of the distribution coefficient with fraction of total nitrogen compounds in the feed is shown by the dashed line and is consistent with the observed variation derived from the boiling point correlation. The phenol distribution coefficients in these same stages could also be correlated to a single value shown by the average line in Figure 3. Subsequent multistage data in the dual solvent system of heptane and dimethylformamide gave a better indication of the variation of the distribution coefficients resembling that obtained for the different boiling ranges.

The most dramatic observation from Figure 4 (d) is the small fraction of the phenolic compounds having distribution coefficients less than unity. Thus 92% of the phenolic compounds have distribution coefficients greater than unity and less than 2% have distribution coefficients below 0.4. It is, therefore, very easy to extract the bulk of the phenolic compounds from the coal liquid compared to the other compounds as shown in Table 2. The values in this table are based upon the assumption that, at a given solvent ratio, compounds with higher distribution coefficients are completely extracted, and none of the compounds with lower distribution coefficients are removed in the solvent. Compensating errors make this a reasonably attainable operating condition for an extraction process. Excessive stages can provide greater degrees of extraction at the given solvent ratio. The relative variations in the degrees of extraction of the different heteroatom compounds are shown graphically in Figure 5.

Solvent ratios will vary depending upon the choice of solvent, and the relative amounts of different type compounds removed at given depths of extraction will also be an indication of the selectivity of the solvent for the different heteroatom compounds.

FRACTIONAL LIQUID EXTRACTION WITH HEPTANE-DIMETHYLFORMAMIDE SYSTEM

Phase separation problems were encountered with the aqueous methanol solvent when the heavy distillate and the high boiling portion of the naphtha fraction were added to the middle distillate. Subsequent work was, therefore, carried out using a dual solvent system of heptane and N, N' dimethylformamide. The DMF proved to be an excellent solvent because it retained in solution the small amount of solids which precipitated at the interface with most other solvents. These solids were suspected to be heteroatom asphaltenes consisting of high molecular weight association compounds of the tar acids and tar bases in the coal liquid. The selection of DMF was initially based on its high solubility for petroleum asphaltenes. It showed an exceptionally rapid phase separation which permitted the operation of an agitated multistage extraction column at higher than normal throughputs and agitator speeds, thereby providing high stage efficiencies.

Preliminary data were obtained for 9-stage countercurrent operation at a solvent ratio of 8 volumes of heptane to 2 volumes of DMF per volume of feed. The previously mentioned composite mixture of the heavy distillate, middle distillate, and a high boiling cut of the naphtha fraction was fed into the center stage of the extraction system.

After phase volumes in the stages became constant and steady state conditions were reached, the liquid phases in all stages were

sampled and weighed. Quantities of the phases in all stages matched the steady state material balances to within 3% with an average discrepancy of less than 1% indicating that the total stream quantities were close to steady state conditions.

Figure 6 shows the comparison of the observed basic nitrogen content of the phases in the stages with the steady state calculations based on 45.6% of the basic nitrogen with a distribution coefficient of 0.6 and the balance with a distribution coefficient of 0.1. These distribution coefficients are expressed as the ratio of weight percent in light phase to weight percent in the heavy phase at equilibrium. Figure 7 shows the same comparison for the phenolic compounds based upon 12% with a distribution coefficient of .2 and the balance with a distribution coefficient of 0.1. The phenol removal from the raffinate was 99.5% under these operating conditions and the basic nitrogen removal was 75%. Reference to the curves in Figure 4 shows that this solvent system has greater selectivity for phenol relative to the basic amines than the previous aqueous methanol solvent.

This solvent system was then studied in a 3" (7.6 cm) diameter 33-stage extraction column with the baffled mixing stage design described by Scheibel (3). Feed was introduced at the eleventh stage above the bottom and different solvent ratios and coal liquid feed rates were used as summarized in Table 3.

The initial objective of the study was to determine the solvent ratio necessary to separate the all heteroatom compounds from the hydrocarbons. Since it was apparent from the aqueous methanol data that complete nitrogen removal would require excessive solvent quantities and that the residual nitrogen compounds were high boiling, the goal was revised to determine the relationship between the solvent ratio and the fraction of the hydrocarbon raffinate which could be distilled to an acceptable nitrogen-free refinery charge stock.

After the first two runs at a ratio of 6 volumes of heptane per volume of DMF and different coal liquid feed rates, a third run was carried out at a heptane ratio of 2.5 to 1 to increase the degree of extraction of the nitrogen compounds. These runs confirmed the difficulty of removing the last traces of nitrogen from the hydrocarbon stream and indicated that hydrotreating may be the most practical process particularly if it were applied to high boiling fractions.

Hydrotreating studies on the raffinates from these extraction runs showed that the rate constant for nitrogen removal was 6 - 7 times the rate constant for the original coal liquid under identical temperature, pressure and catalyst conditions. The extraction work was then directed toward evaluating the minimum amount of extract product that could be removed to give the same high denitrogenation rate. The fourth run on the extraction column was carried out at a heptane to DMF ratio of 8. About 16% of the coal liquid was extracted into the DMF and the denitrogenation rate during hydrotreating was the same as with the higher degrees of extraction.

CONCLUSIONS

The distribution coefficients for the different types of heteroatom compounds in coal liquids decrease with increasing boiling point. Phenolic compounds are the easiest to extract and basic nitrogen compounds are the most difficult. The selectivity of DMF for phenolic compounds relative to the nitrogen compounds is greater than the selectivity of aqueous methanol.

Extraction of the coal liquids to remove about 15% of the feed in the solvent results in a raffinate that can be hydrotreated to reduce the nitrogen concentration at a rate which is 6 - 7 times the rate obtained for the original coal liquid. The reasons for this improvement are not entirely clear and additional work is in progress to identify the cause. The results of the hydrogenation studies will be reported at a future date when they are completed. Preliminary data indicate that the milder hydrotreatment of the raffinate eliminates the nitrogen with less hydrogenation of the aromatic compounds and yields an excellent catalytic cracking charge stock.

The present work does not presume to present the most economic approach to converting coal liquids to gasoline. The results demonstrate only the potential for liquid extraction in the production of synfuel from coal. It will be necessary to identify the heteroatom compounds which inhibit the hydrogenation and denitrogenation of the coal liquid raffinate before the most practical solvent system can be established and the optimum process design developed.

The substantial benefits derived from fractional liquid extraction of coal liquids can accelerate the commercialization of this synfuel process and reduce the time required for this country to once again become energy sufficient.

ACKNOWLEDGEMENT

This work was carried out under DOE Contract EF-76-C-01-2306.

LITERATURE CITED

(1) Scheibel, E. G., Ind. Eng. Chem., 49, 1679 (1957).

(2) Scheibel, E. G., Ind. Eng. Chem., 44, 2942, (1952)

(3) Scheibel, E. G., AIChE Journal, 2, 74 (1956).

SCHEIBEL, E.G.   SCHNEIDER, A.   HOLLSTEIN, E.

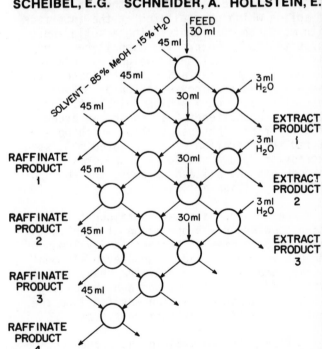

Figure 1. Development of a four-stage countercurrent extraction.

Table 1a. Summary of distillation and extraction data for SRC-II liquid fractions.

Cut No.	Temp Range °C	Wt.% Total N	Wt.% Basic N	Wt.% Total O	Wt.% Phenol	Product Cycle	Wt.% Total N	Wt.% Basic N	Wt.% Total O	Wt.% Phenol
1	74-190	.62	.42	5.41	26.6	RP-3	.0151	.012	.03	.01
						RP-4	.0109	.048	.10	.06
2	100-200	.92	.71	5.88	28.0	RP-3	.0276	.0045	.03	.02
						RP-4	.0172	.0040	.02	.05
3	200-213	1.49	1.29	4.72	13.6	RP-3	.028	.0095	.08	.05
						RP-4	.035	.0160	.05	.04
4	213-219	.96	.77	2.19	10.3	Fraction crystallized at room temperature - not extracted				
5	219-234	1.05	.90	3.80	5.7	RP-3	.12	.091	.090	.10
						RP-4	.15	.107	.12	.13
6	234-242	.95	.72	2.30	2.95	RP-4	.27	.320	.43	.42
						RP-4	.30	.357	.41	.54
7	242-256	1.39	1.03	2.88	2.4	RP-3	.55	.442	.76	.54
						RP-4	.63	.510	.93	.67
8	256-270	1.62	.90	2.36	.7	RP-3	.78	.591	1.16	.49
						RP-4	.88	.636	1.16	.55
9	270-293	1.04	.72	1.73	.52	RP-3	.72	.538		.32
						RP-4	.59	.570	1.26	.40
10	239 +	.87	.56	1.38	.42	RP-3	.71	.47	.98	.105
						RP-4	.73	.49		.12
Naphtha Cut										
10	176-177	.70	.45	6.39	32.4	RP-3	.0153	.00095	.05	.01
						RP-4	.0166	.0047	.03	.06
Total Naphtha		.52	.21	2.04	9.04	RP-3	.0436	.0135	.07	.04
						RP-4	.0563	.0148	.04	.03
Heavy Distillate										
Total		1.36	.65	2.33	.89	RP-3	1.09	.64	1.82	.040
						RP-4	.94	.56	1.72	.023

31

Table 1b. Summary of average distribution coefficients evaluated from extraction data for SRC-II liquid distillate fractions.

Middle Distillate Cuts

Cut No.	Solvent Ratio L/H	Product Cycle	Fractions Unextracted				Distribution Coefficients			
			Basic N	Total N	Phenol	Total O	Basic N	Total N	Phenol	Total O
1	3.44	HP-3	.00115	.0095	.00015	.0022	2.73	1.27	5.47	2.15
		HP-4	.0046	.0068	.00091	.0074	1.65	1.43	2.97	1.39
2	4.99	HP-3	.00166	.0079	.00019	.00134	1.62	.94	3.45	1.76
		HP-4	.00147	.0049	.00049	.00089	1.72	1.11	2.36	2.02
3	5.50	HP-3	.00174	.0044	.00086	.00405	1.45	1.04	1.87	1.07
		HP-4	.00296	.0055	.00068	.00243	1.20	.97	2.02	1.29
5	3.52	HP-3	.0378	.044	.0068	.0088	.72	.67	1.34	1.28
		HP-4	.0443	.038	.0085	.0117	.68	.72	1.30	1.13
6	2.00	HP-3	.277	.182	.092	.120	.46	.55	.86	.75
		HP-4	.319	.204	.118	.115	.43	.56	.77	.77
7	2.16	HP-3	.252	.233	.131	.157	.46	.48	.66	.61
		HP-4	.291	.268	.164	.192	.42	.44	.60	.55
8	1.76	HP-3	.472	.345	.502	.353	.34	.45	.31	.44
		HP-4	.510	.390	.563	.353	.31	.41	.27	.44
9	1.55	HP-3	.608	.565	.503		.27	.30	.35	
		HP-4	.645	.465	.629	.592	.25	.39	.25	.27
10	1.29	HP-3	.808	.626	.241	.685	.156	.31	.80	.17
		HP-4	.842	.645	.275		.125	.29	.74	

Naphtha Cuts

Cut No.	Solvent Ratio L/H	Product Cycle	Basic N	Total N	Phenol	Total O	Basic N	Total N	Phenol	Total O
10	2.95	HP-3	.0019	.0097	.000134	.0039	2.64	1.47	6.5	1.90
		HP-4	.0021	.0105	.00082	.0026	2.57	1.43	3.5	2.34
Total	2.29	HP-3	.0436	.057	.00316	.0230	1.50	.94	2.84	1.35
		HP-4	.0476	.074	.00226	.0131	1.03	.84	3.1	1.69

Heavy Distillate

Cut No.	Solvent Ratio L/H	Product Cycle	Basic N	Total N	Phenol	Total O	Basic N	Total N	Phenol	Total O
Total	1.28	HP-3	.852	.716	.40	.774	.12	.23	.55	.19
		HP-4	.746	.617	.23	.715	.20	.32	.83	.23

Table 2. Fractions of different hetero-atom compounds extractable at varying solvent ratios.

Solvent Ratio	Percent of Compounds Removed			
	Phenolic	Total Oxygen	Total Nitrogen	Basic Nitrogen
1.0	92.0	64	21	33.0
2.0	97.4	83	55	54.0
3.0	99.0	91	92	82.0
5.0	99.8	98	100	93.0
10.0	100.0	100	100	99.5

Table 3. Extraction of SRC-II composite with heptane-DMF solvent system in 33-stage baffled extraction column.

Run No.	1	2	3	4
Volumetric Ratios				
Heptane	12.0	12.0	5.0	12.0
DMF	2.0	2.0	2.0	1.5
Feed	1.0	1.5	1.0	1.0
Fraction Extracted				
Wt % of Feed	27.0	25.0	60.0	16.50
Feed Composition				
Wt % Total N	1.19	1.19	1.19	1.19
Wt % Total O	3.86	3.86	3.86	3.86
Wt % Basic N	0.84	.84	.84	.84
Wt % Phenol	7.60	7.60	7.60	7.60
Raffinate Product				
Wt % Total N	.38	.39	0.22	0.50
Wt % Total O	.31	.23	0.27	0.44
Wt % Basic N	.35	.36	0.24	0.45
Wt % Phenol	0.10	0.40	0.0	0.02
Extract Product				
Wt % Total N	2.46	2.70	1.99	4.49
Wt % Total O	8.11	7.79	6.75	11.14
Wt % Basic N	.99	1.00	0.82	0.80
Wt % Phenol	13.70	11.90	9.90	17.00

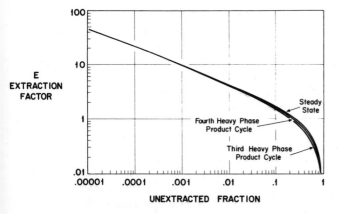

Figure 2. Extraction factor vs. unextracted fraction at different product cycles.

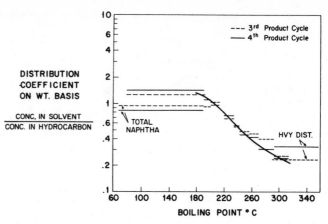

Figure 3a. Distribution coefficients of total nitrogen in different boiling fractions.

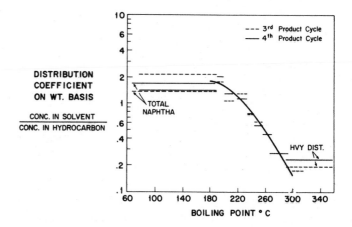

Figure 3b. Distribution coefficients of total oxygen in different boiling fractions.

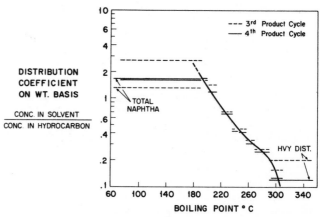

Figure 3c. Distribution coefficients of basic nitrogen in different boiling fractions.

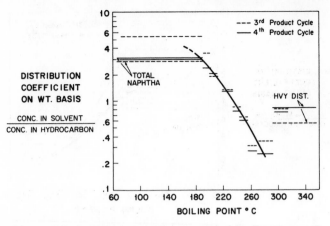

Figure 3d. Distribution coefficients of phenolics in different boiling fractions.

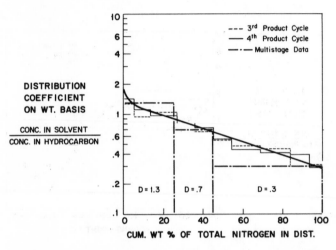

Figure 4a. Total nitrogen distribution coefficient variation with fraction distilled.

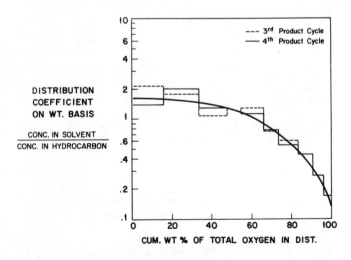

Figure 4b. Total oxygen distribution coefficient variation with fraction distilled.

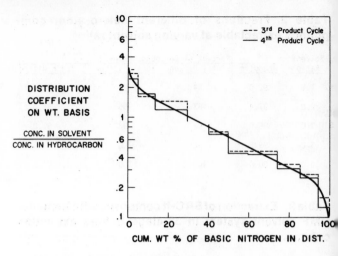

Figure 4c. Basic nitrogen distribution coefficient variation with fraction distilled.

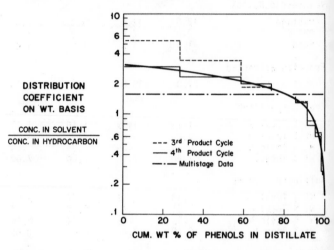

Figure 4d. Basic nitrogen coefficient variaion with fraction distilled.

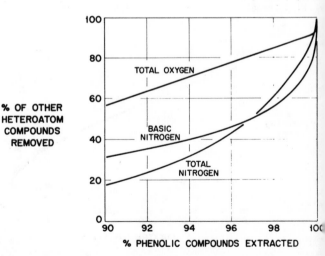

Figure 5. Relationship between phenol extraction and removal of other hetero-atom compounds.

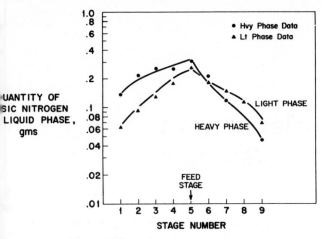

Figure 6. Basic nitrogen in nine-stage extraction system.

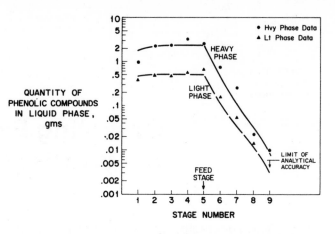

Figure 7. Phenols in nine-stage extraction system.

Operating Experience with the Westinghouse Gasifier

The successful operation at Westinghouse gasification process, under development since 1972, has been achieved by developing reliable techniques, for startup, shutdown, and full-load operation.

E.J. Chelen, D. Revay, and P. Cherish
Westinghouse, Electric Corp.
Madison, Pa.

Many promising developmental process concepts are never shown to be feasible simply because of the inability to start up and operate test units successfully so that process data can be developed. In the development of process technology utilizing reduced-scale equipment such as a process developmental unit (PDU), an equally important and parallel effort must be considered to gain the full benefits of the development effort. Included in this effort is the development of operational techniques and the design of auxiliary components and subsystems. The successful operation of the Westinghouse process development unit over the past five years has been achieved through a combined process development-component design approach coupled with a test management discipline, including careful planning, control and execution. These are summarized in the following sections. The result of the integrated approach is the successful operation of the PDU on a nominal 50 percent downtime basis. In the 7000 hours of hot operation since 1975, a substantial data base has been achieved for scaling the process and the related hardware to a commercial size.

TYPICAL EXPERIMENTAL TEST OPERATION

To start up the plant with a minimum of leaks and nuisance anomalies, which would prohibit achieving steady-state conditions, a series of procedures and checks are performed at the PDU. Figure 1 depicts the sequence of events involved in starting up and operating the unit.

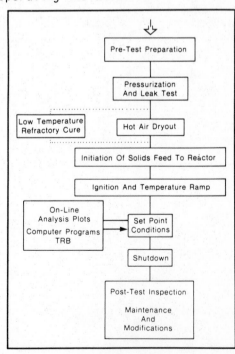

Figure 1. Sequence of PDU test events

Figure 2 shows one of the major subsystems involved and some typical hardware components.

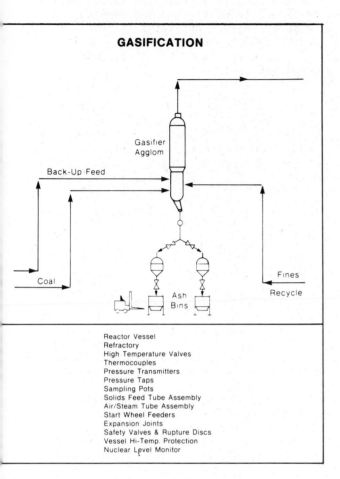

GASIFICATION

Back-Up Feed

Gasifier Agglom

Coal

Fines

Recycle

Ash Bins

Reactor Vessel
Refractory
High Temperature Valves
Thermocouples
Pressure Transmitters
Pressure Taps
Sampling Pots
Solids Feed Tube Assembly
Air/Steam Tube Assembly
Start Wheel Feeders
Expansion Joints
Safety Valves & Rupture Discs
Vessel Hi-Temp. Protection
Nuclear Level Monitor

Figure 2. Typical PDU components

During the pretest preparation phase of the test, the test plan document and the master test procedure are prepared. The test plan defines the objectives of the test and outlines the set point conditions to achieve those objectives. Included in this document is a failure mode analysis of the system and a failure action plan, which defines the steps to be followed in the event of a failure. The master test procedure elaborates on the test plan. It details a step-by-step procedure for attaining the test goals. Each step must be signed off before proceeding with the next step of the procedure. Both of these documents are reviewed and approved by operations and engineering groups prior to initiation of the test. Procedures for the various subsystems are also reviewed and revised, as necessary, prior to the test.

During this phase, the feedstock inventory is checked and lab analysis performed to ensure that the size cut and chemical analysis meet specifications. Also, delivery of utility gases required for the test, such as carbon dioxide, propane, and oxygen, is scheduled. Pretest briefings are held among operating crews and shift engineers to discuss the goals of the test, set point conditions, sampling schedule, and modifications implemented for the test. If a new system or modification is installed, special training sessions are scheduled to familiarize people with the details of the modification, the operating procedure, the potential hazards, and the procedure to follow should a failure occur. A pretest checklist procedure is also implemented. This involves steps such as stroking valves, calibrating critical flow transmitters, and ensuring that all instrumentation needed for the test is installed and properly calibrated.

The plant is then pressurized in 50 psig (0.345 MPa) increments to 200 psig (1.378 MPa) to detect leaks on major vessel flanges and on the various sample trains used during the test. After repairs are made on all detectable leaks, the system is pressurized to 250 psig (1.724 MPa) and locked. A pressure leak rate is then determined. This leak rate must be equal to or less than the standard acceptable leak rate of 30 psig (0.207 MPa) per hour before the test is continued into the next phase.

Next, a refractory hot air dryout of the system is conducted by increasing the temperature 50°F (10°C) per hour until 500°F (260°C) is achieved in the freeboard section of the reactor. This temperature is held for approximately 12 hours. During this soak period, the ash lockhoppers are alternately vented to the atmosphere to prevent moisture from condensing in the lockhoppers. The solids transport line from the cyclone to the lockhoppers is vented to the atmosphere to prevent moisture from condensing in the bottom section of the cyclone. However, if a new section of refractory was installed, a curing cycle is also performed. The heatup rate for the curing cycle is still 50°F (10°C) per hour, but an 8-hour hold period is initiated when the reactor temperature reaches 250°F (121°C). This is to drive off the moisture in the refractory prior to continuing the temperature ramp to 500°F (260°C) for the 12-hour hold.

After the hot air dryout is completed, preparations are made to initiate solids feed to the reactor. The load cells on the empty lockhoppers are "zeroed," and coke breeze is transferred from the storage bins to the lockhopper. The recycle gas flows are started in the transport gas lines, and line velocities are checked to ensure that the minimum carrying velocity requirement is met. Low flow switches tie into the transport gas flow transmitters to protect the recycle gas heater elements. Computer alarm programs alert the control room operators of low flow conditions or of a high pressure drop across the transport line.

Coke breeze feedstock is then introduced into the reactor and the bed level is built to 25 feet (7.62 m). The bed level is monitored with a nuclear densitometer and also with differential pressure transmitters. Once the bed is built, the process air temperature is slowly raised from 500°F (260°C) to a maximum of 1000°F (538°C). Combustion of coke breeze usually occurs when the reactor bed temperature is in the range of 650 to 700°F (343.5 to 371°C). The critical flows and temperatures are monitored by a "trend" program on the computer and by CRT displays. Immediately after ignition, the air flow rate is reduced to a minimum, and a small flow of carbon dioxide is introduced through the air tube to maintain a velocity of greater than 40 feet per second (12.19 meter/second). The temperature ramp is controlled by slowly backing out the carbon dioxide flow and slowly increasing the air flow. A 100 to 150°F (38 to 65.5°C) per-hour heat rate is implemented. As the temperature in the reactor approaches 1600°F (871°C), a 4-hour hold is initiated. This is to allow the reactor to approach a thermal equilibrium. After the 4-hour hold period, coal feed to the reactor is initiated and the temperature ramp continues to set point conditions.

Set point conditions are declared when the temperature and flow conditions specified in the test plan are achieved. During the test, log books are maintained to record anomalies and daily activities. Operation checklists are used once per work shift to monitor vessel wall temperatures. air quality samples, and equipment parameters. A sample schedule ensures that all samples needed for data reduction are obtained. Throughout the test, plots of the ash concentration and the bulk density of the solid samples versus time are generated. Various computer programs are also on line to calculate parameters such as air tube velocity, feedline velocity, freeboard velocity, and bed density. A "report program" generates an hourly average of critical reactor parameters and is used as a quick reference for steady-state conditions.

Every day of the test, a Technical Review Board (TRB) meeting is conducted in the morning to review the test status. The members of the board, as a minimum, consist of the Operations Manager or his assigned alternate, Shift Test Supervisor, Test Coordinator and Shift Engineer. The following agenda items are discussed:

- Summary of plant status during the past 12 hours

- Safety items

- Operational problems

- Current set point conditions

- Conditions and parameters that must be met for the next 12 hours.

As a result of the TRB meeting, a 12-hour test plan may be generated by the Test Coordinator and Shift Engineer to change the set point conditions from those originally defined in the master test procedure.

After the goals of the test are achieved, a shutdown is initiated. The reactor temperatures are slowly reduced by decreasing the air flow rate. A 16- to 18-foot (4.88 to 5.49 m) bed level is maintained until combustion is quenched. Once the reactor temperatures fall below 1200°F (649°C), depressurization of the system begins at a rate of 30 psig (0.207 MPa) per hour. When the reactor bed is completely withdrawn and the temperatures fall below 300°F (149°C), all flows to the reactor are shut off and the vessel is allowed to cool by natural convection.

Upon conclusion of a test, disassembly procedures are initiated, photographs are taken and inspection reports are written. After reviewing the inspection reports, a post-test procedure is implemented, which consists of cleaning, replacing filters, taking thickness measurements of piping, repairing damaged equipment, and conducting routine mechanical equipment and instrumentation maintenance and calibrations. Post-test debriefing meetings are held among operating crews and shift engineers to discuss test data, test anomalies, and future plans and modifications.

After the test, the data are reviewed in detail and steady-state test points are declared. A steady state is defined as that period during the set point in which all flow and pressure parameters fall within ± 5 percent of the average value and all temperatures hold to ± 2.5 percent of the average value. These rigorous criteria are used to ensure that the determination of steady state is valid. A heat and material balance is performed on the as-measured data; no forcing of data is performed. The test data and heat and material balances are compiled into a final test report issued by the Test Coordinator.

TYPICAL TEST RESULTS

Normally, the above sequence consumes about two weeks and produces about 200 hours of steady-state process data. The data base for scale-up, firmly built on the basis of several 24- to 48-hour duration steady-state periods during the course of each test, encompasses a wide range of parameters. Longer-duration tests at selected "design points" are now being emphasized to ensure that long-term effects are accounted for in the overall process evaluation. Table 1 presents a comparison between PDU test results and commercial plant predicted operation.

It will never be possible on a small PDU scale to achieve commercial size design point operation because of geometry and heat load limitations. Heat losses from reaction equipment are from 5 to 10 percent for the PDU, as compared with an estimated 0.6 percent for a commercial-scale design. Recycle gas usage also contributes 20 percent additional heat load on the PDU and only 1 percent on the scale-up plant. However, every attempt is being made to optimize PDU performance and to approach, as closely as possible, commercial design point operation. In particular, several heat conserving techniques are being implemented to minimize recycle gas use. These include air transport, dense phase transport of feed coal, and minimizing the use of fluidizing gas in the ash annulus.

The experimental data from the PDU must serve two seemingly conflicting requirements. First, it should be extensive in terms of covering a broad canvas of scale-up design and operational issues. Second, it must be intensive enough to provide a thorough understanding of, and a solution to, each design problem. The experimental and operational program is directed at achieving both of these goals and has done so adequately. Broad problems have also been solved in the process of maintaining the plant as a reliable operating unit.

OPERATING HARDWARE EXPERIENCE

From the experience gained in operating the plant in the last five years, many modifications to the system have been incorporated. Some of the key hardware design achievements of the Westinghouse process are summarized in Table 2.

Table 1. Typical PDU test results versus commercial plant operation.

| | Air Blown | | Oxygen Blown | |
	Commercial	PDU	Commercial	PDU
Coal Type	Illinois No. 6	Indiana No. 7	Pittsburgh No. 8	Ohio No. 9
Superficial Bed Velocity, fps	2.5	2.13	2.5	1.86
Freeboard Temperature, °F	1850	1828	1850	1800
Steam/Coal (MAF)	0.25	0.30	0.50	1.06
Recycle Gas/Lb Coal	0.20	1.45	0.50	1.28
CO_2 Purge/Lb Coal	0.0004	0.17	0.004	0.23
Coal Feed Rate, Lb/Hr	65,000	1070	73,500	858
	(6% moist)	(11.9% moist)		(4% moist)
Net Gas Output, scfh	4.132×10^6	3.8699×10^4	2.776×10^6	3.312×10^4
System Pressure, psig	340	225	340	130
HHV (Dry Basis) Raw Gas, Btu/scf	155	100	346	260
Product Gas Composition				
Carbon Monoxide	26.05	17.52	51.39	36.7
Hydrogen	14.89	11.91	29.38	36.4
Carbon Dioxide	4.87	9.31	9.37	23.6
Nitrogen	50.82	60.71	0.36	–
Methane	2.24	0.54	8.36	2.4
Hydrogen Sulfide	0.72	NEG	0.36	0.8

Table 2. Hardware designed developed and demonstrated.

- Draft tube
- Conical grid plates
- Coal injection
- Oxidant injection
- Refractory linings
- Internal geometries
- Lockhoppers
- Cyclone internals
- Gas distributors

- Feed transport valves
- Feed lines
- Char transfer leg
- Water quench and scrubbing
- Waste processing
- Slurry letdown valves
- Gas characterization instruments
- High temperature instrumentation
- Sampling apparatus

During the design of the PDU, three state-of-the-art solids flow measuring techniques were selected to measure material feed to the reactor. The three methods consist of belt weighers, speed-controlled starwheel feeders, and lockhoppers with strain gage load cells. These systems are quite reliable, but to achieve this reliability, modifications were implemented.

During plant startup, problems were noted with the lockhopper load cell installation. As the lockhopper vessels were pressurized, both axial and transverse stresses were imposed on the load cells from the rigid vent piping and mismatched expansion joints between the vessels. These problems added a false weight to the load cells and, in some cases, over-ranged the instrumentation. Most of the problems were corrected by eliminating the mechanical restraints in an attempt to make the hoppers free floating regardless of forces generated by internal pressure or external forces. Matching expansion joints were installed in critical areas above and beneath each lockhopper, in addition to flex lines in connecting piping. A modification was also made in the tare adjustment controls to permit a wider range of tare adjustment during the startup phase. In practice, the load cells are calibrated at atmospheric pressure with no material in the hoppers. When the lockhoppers are pressurized to system pressure, the tare control is adjusted to show an indicated weight of zero in the control room. Filling of the lockhopper and continuous solids monitoring is then accomplished using the new zero offset.

Starwheel feeders are used for solids flow rate control. The major problem with the starwheel feeders is the uncertainty in the amount of material that the feeder handles. Although the feeder revolution rate and the feeder pocket volume are known, the pockets are not uniformly filled. Particle size, moisture content, and other variables affect the feeder mass transfer rate for a given revolution rate. In an effort to improve on these feeders, carbon dioxide purges were installed on the inspection plates of the units. This purge gas is used when it is suspected that the pockets may be packed with wet material. At the present state of development, this feed system can control and measure the solids feed rate to within 5 percent. In conjunction with the above, belt weighers

are used to measure batches of feedstock material from storage bins to feed lockhoppers. However, if transfer rates are not matched properly or if the material is wet, problems with the belt staying on the track will occur. It appears that this unit requires too much maintenance during operation, and evaluations are being conducted to eliminate this unit at the PDU.

In an effort to obtain uninterrupted coal feed to the gasifier, a number of revisions were made to the pneumatic conveying lines, the results of which are more than satisfactory. For instance, if feed is interrupted for some reason, a backup system maintains steady-state conditions in the reactor. Block and bleed valves isolate the feedline and associated lockhopper system from the rest of the process. The isolated feed system can then be worked on under ambient temperature and atmospheric pressure conditions without disturbing gasifier parameters. Operating maps define critical transport velocities in the pneumatic conveying lines, and flow and temperature alarms warn operators of a potential low velocity problem, preventing feed interruption to the gasifier. A unique erosion-resistant feed line elbow design minimizes erosion. Studies are also being made to determine erosion rates of various materials tested in this system.

During earlier tests, many problems were encountered with the as-built thermocouple and thermowell assemblies. The original design consisted of a Hastalloy X thermowell and a W5 thermocouple (tungsten rhenium, grounded junction, non-compacted MgO insulator in a disilicide-coated moly sheath). The thermowells were welded to reactor flanges and varied in length, thus requiring a stock of different size thermocouples for each well size.

This restraint imposed a spare parts problem since there is no standard size thermocouple unit to fit any given thermowell. Also, there is no pressure seal between the thermocouple and well to keep hot gas from exiting the reactor in the event of a thermowell failure. The thermocouple assemblies failed due to degradation from hydrogen sulfide attack, severe temperature cycling, and oxidation. To alleviate the various failure mode problems, a new thermocouple and thermowell assembly design evolved utilizing off-the-shelf components. This unit was designed to incorporate a "change-on-line" feature. The design highlights are: (1) use of an ungrounded junction W5 thermocouple to eliminate wire failure from differential thermal expansion between wire and sheath; (2) use of 310 stainless steel thermowells instead of Hastalloy X; and (3) use of a block valve and pressure gage for safe removal of a thermocouple on line. The pressure gage indicates the condition of the well and the block valve isolates the assembly when the thermocouple is removed through a pressure seal fitting. Satisfactory performance has been obtained from the new design.

Maintaining consistent operation of the recycle gas and process air compressors was another problem to be solved. Maintenance, repair, and overhauling of the compressors were done after nearly every test run of the gasifier. To improve on the efficiency and dependability of these units, the following modifications were implemented:

- An in-line filter was installed on the suction side of the process air compressors to prevent rust, scale and dirt from entering the units.

- A temperature controller was installed on the glycol cooling loop for the process air compressor jackets to guard against saturation temperatures.

- A refrigeration unit was placed between the series process air compressors to prevent damage to the Teflon piston rings in the second reciprocating compressor as a result of moisture condensation.

- A 10-micron cloth filter unit was installed on the suction side of the recycle gas compressor to minimize excessive wear and plugging of the intake valves, which results from char fines in the gas stream.

After the modifications, estimated downtime went from once every 200 hours of operation to once every 1000 hours of hot operation.

The product gas quench and scrubbing systems and the liquid and waste solids handling systems were also modified to extend operational service time and improve the reliability of these systems. New pump casing coatings, the controlled addition of flocculant, and pH adjustment additives and corrosion inhibitors significantly improved the operation time of this subsystem.

Furthermore, parallel letdown valves, with special seats and plated balls, allow the process to remain on stream while routine maintenance is performed on the alternate valves and the valve actuators. With the recent installation of the waste solids separator, waste solids removal and recycling of the fines collected is now accomplished at the PDU. Also, the disposal of scrub liquor and waste sludge in a safe and environmentally acceptable manner has been demonstrated.

During the past five years, new refractory was cast for the cyclones, the new expanded vessel sections, and various interconnecting piping sections, with no major problems experienced with spalling, cracking, or erosion.

Techniques developed and used successfully in the PDU are summarized below:

Casting Procedure

Control procedures ensure consistency during the mixing and continuous pouring cycles and provide data for the development of the latter techniques. The consistency of the vibration techniques used during a casting are crucial to the success of this work.

Cure Cycle

The temperature ramps for heating the refractory and driving the moisture off at $212^{\circ}F$ ($100^{\circ}C$) appear to be very critical to avoid producing cracks in the refractory. The heating and cooling cycles are controlled up to $500^{\circ}F$ ($260^{\circ}C$). A temperature hold is maintained for approximately 12 hours throughout the system to ensure that most of the moisture is driven off.

Because of the two-layer castable refractory construction used in the PDU, nominal pressurization and depressurization rates are a maximum of 50 psig (0.345 MPa) per hour and 75 psig (0.517 MPa) per hour, respectively. No adverse effects on the structural integrity of the refractory has been observed in the five years of operation. At the PDU, these rates are probably very conservative and future testing may be directed toward investigating this area more closely to study transient effects.

In support of transient testing, a computer program provides supervisory pressure control of the PDU. The program, which controls the set point on a pressure control indicator in order to pressurize or depressurize the gasifier, was successfully demonstrated after overcoming a few initial problems. Achieving this milestone opens the door to exploring other areas of the process that can possibly be computer controlled.

In the PDU, the process air and recycle gas to the reactor is heated by electrical gas heaters. Arcing of the electrical feed, through experience in earlier series of tests, was a constant maintenance problem. A unique insulator design modification, using a compressed Teflon sleeve, virtually eliminated the arcing problem, which was the result of wet conductive material formed by these gas streams. In addition, selectively located block valves downstream of the heaters prevent back-flow of carbonaceous fines into the heaters during startup.

Also, several modifications to the gasifier improved operational and process performance. A new conical grid design provides a more efficient distribution of the gas and promotes more solids circulation in the area. Several oxidant injectors were designed and tested in this unit. Reactor length-to-diameter (L/D) ratios were changed and diameter ratios were altered as a result of the process and corresponding operational data obtained from the PDU. These modifications are typical examples of the spinoff technology that evolves with the development of any process. Most of this spinoff knowledge and experience gained during the operation of a Westinghouse-size PDU can be applied as a data base for the scale-up design and operation of a commercial-size plant.

CONCLUSION

The experimental program and approach employed by Westinghouse to operate the process development unit at Waltz Mill, Pennsylvania, is designed to provide a sufficient data base for the scale-up of the process to a commercial size. In addition,

data on operational techniques, component and subsystem design, and operational and equipment reliability are obtained to guide the plant designer and operators of the commercial version of the process. These data are acquired through a disciplined approach to experimental design, test operation, design review, and documentation to make efficient use of material, time, and human resources and to achieve a practical and meaningful data base.

CHELEN, E.J.

Upgrading Solvent-Refined Coal by Hydrogenation

It is possible to convert the solvent-refined coal to lighter and cleaner fuel and to transportation fuels by deep hydrogenation.

J. Caspers, and R.P. Van Driesen
C-E Lummus Combustion Engineering, Inc.
Bloomfield, N.J.

K. Hastings
Cities Service
Tulsa, Ok.

S. Morris
International Coal Refining Co.
Allentown, Pa.

Solvent refining of coal followed by ash removal produces a relatively clean burning solid fuel. The material is fluid and pumpable at temperatures above 350°F (177°C) and could conceivably be used as a liquid fuel at preheat temperatures of 500-600°F (260-316°C). It is possible to convert the solvent refined coal to lighter and cleaner fuel and to transportation fuels by deep hydrogenation.

Pilot plant work on the upgrading of solvent refined coal by hydrogenation has been performed in The Lummus Company's New Brunswick, New Jersey laboratories almost continuously for four years. This work which utilizes the LC-Fining[SM] process has been described in several prior publications. In the last eighteen months, studies and designs have been made for commerical operation. The LC-Fining process will be included in the 6,000 ton per day (5443 Mg/d) SRC-I demonstration plant to be installed at Newman, Kentucky, and operated by the International Coal Refining Co., a joint venture of Air Products & Chemicals, Inc. and Wheelabrator-Frye, Inc.

In mid-1979, a Phase Zero design study was made on the process. Design basis was 2289 tons per day (2077 Mg/d) of solvent refined coal feed to be converted into lighter products at a conversion rate of 50 volume percent to 850°F (454°C) and lighter material. A simplified process flow sheet is shown in Figure 1. Preheated solvent refined coal is mixed with a solvent and the mixture pumped into a process heater. Hydrogen is passed through a separate process heater and the two fluid streams are mixed and flow into the first of two reactors. The mixed liquid and gas stream comes into the bottom of the reactor at a temperature somewhat lower than the reactor temperature. The heat of reaction brings the relatively cool feed material up to reactor temperature. This occurs rapidly for two reasons. There is a recycle of liquid from the top of the reactor to the bottom at rates several times the fresh feed rate. Also there is a top to bottom turbulence and mixing within the reactor. The

*LC-Fining is a service mark of C-E Lummus for engineering, marketing and technical services relating to hydrocracking and hydrodesulfurization processes for reduced crude and residual oils.

reactor operates essentially isothermal throughout.

The entire effluent stream passes from the first reactor to the second. Because the second reactor operates at the same temperature as the first, it is necessary to decrease the enthalpy of the first reactor effluent relative to the second reactor temperature by an amount equal to the second reactor heat of reaction. This is done by quenching the first reactor effluent with a suitable amount of relatively cool hydrogen rich gas.

The reactors operate with the catalyst in an expanded or fluidized state. This is accomplished by the upward flow of liquid through the reactor which expands the catalyst bed to occupy about one third more volume than the settled bed. Fresh catalyst is added to and spent catalyst is withdrawn from the reactors during the operation.

This compensates for catalyst activity loss due to carbon deposition and a steady state or "equilibrium" average catalyst activity is established for the reactor bed. This results in a constant product quality when feedstock and catalyst replacement rate are maintained constant or the opportunity to control product quality by adjusting catalyst replacement rate. There is no need to interrupt operation for catalyst change out.

Effluent from the second reactor passes to a separator at pressure and temperature conditions identical to the reactor. Liquid is withdrawn from the separator and let down in pressure before going to atmospheric and vacuum fractionation. Vapor effluent is cooled and condensates removed leaving a relatively hydrogen rich gas stream which is further upgraded in hydrogen content by first removing some of the light hydrocarbons and then adding fresh hydrogen. The hydrogen rich stream is heated by exchange with effluent and finally in the furnace before mixing with the liquid feed and passing into the first reactor. Light hydrocarbon removal from the hydrogen recycle stream is accomplished by scrubbing with a lean oil which preferentially dissolves the hydrocarbons.

Feed and product qualities are shown in Table 1. These have been adjusted slightly to coincide with a more recent feed analysis. Feedstock is of an extremely low API gravity and relatively high nitrogen content. Product gravities are much improved though

still lower than petroleum streams. The naptha produced should make an excellent feedstock for reforming to high octane gasoline blend stock after some treatment for nitrogen removal. Heavier distillates could be blended into distillate fuels used as cutter stock for residual fuel. The unconverted heavy material boiling above 850°F (454°C) could be used as boiler fuel much as the solvent refined coal is. It is somewhat cleaner than the solvent refined coal in that the sulfur and nitrogen contents are lower. Also it is expected that this unconverted 850°F+ (454°C+) material would be more suitable for cutting back and handling as heavy liquid fuel.

Conversion to distillates is about 50 percent complete for this operation. It may be desirable to convert much higher quantities to distillates. Shown in Table 2 are estimated yields for converting about 85 percent of the solvent refined coal to distillates. Naphtha range yield is 48.2 volume percent on feed compared to 20.3 volume percent on feed for the 50 percent conversion case. Hydrogen consumptions are 9400 SCF per ton (293 cubic meters at 101.3 kPa and 15 degrees Celsius per megagram) for 50 percent conversion and 22000 SCF per ton (687 cubic meters at 101.3 kPa and 15 degrees Celsius per megagram) for 85 percent conversion.

It would be possible to accomplish either the 50 percent conversion operation of the 85 percent conversion in the same unit. For instance, the unit designed for 50 percent conversion for the Phase Zero design could be utilized for 85 percent conversion of 1145 tons per day (1038 Mg/d) of solvent refined coal with very moderate changes in the downstream handling equipment and a slight increase in hydrogen consumption.

Catalyst consumption will be about 1.4 pounds per ton (0.7 kg/Mg) for the 50 percent conversion case and 2.8 pounds per ton (1.4 kg/Mg) for the 85 percent conversion case. In a unit used to perform either case the catalyst consumption per day would be the same over the conversion range 50 percent to 85 percent.

Estimated investments and utilities required for the 2289 ton per day (2077 Mg/d) unit are shown in Table 3. The investment is on a January 1980 basis. Included are all inside battery limits equipment for LC-Fining. Also included in Table 3 are estimated investment costs for an LC-Fining facility to process 17,168 tons per day (15 574 Mg/d) of solvent refined coal at conversion levels of 50 percent. This would correspond to coal feed to the **SRC-1** plant of 30,000 tons per day (27 216 Mg/d). Three separate LC-Fining trains would be used in this large commercial design at a 50 percent conversion. The plant designed for 50 percent conversion could be used at 85 percent conversion with throughput reduced to 8588 tons per day (7790 Mg/d) of solvent refined coal. Investment for such a plant is estimated to be 385 million dollars on a 1980 basis.

A plant of this size could produce 44,800 BPSD (7123 m^3/d) of distillates for refinery feed boiling in the range C_4-850°F (454°C). Of this total, 16,200 BPSD (2576 m^3/d) would boil in the range of C_4-390°F (199°C), suitable for gasoline blending or catalytic reforming. There would be 8205 tons per day (7443 (Mg/d) of low sulfur 850°F+ (454°C+) material suitable for use as a clean boiler fuel.

When operating this size plant at 8588 tons per day (7790 Mg/d) feed and an 85 percent conversion the yield of naphtha would be 22,400 BPD (3561 m^3/d), about 6000 BPD (.953 m^3/d) greater than for the 50 percent conversion case. Overall yield of C_4-850°F (454°C) distillate would be less at the higher conversion, 41,100 BPSD (6534 m^3/d) versus 44,800 BPSD (7123 m^3/d). Also hydrogen consumption would be greater, 188 million SCFD versus 161 million SCFD. Overall it is felt that such an operation may at times prove attractive by reason of the higher naptha yield.

A plant provided for high conversion could produce 82,300 BPSD (13 084 m^3/d) of distillates of which 38,400 BPSD (5533 m^3/d) would boil in the C_4-390°F (199°C) range. In fact the yield of distillate from such a plant is of greater volume than the solvent refined coal feed. No figures for such a plant are shown in Table 3. The investment would be approximately twice that shown since duplicate trains would be used. Utilities would be less than twice those shown. In fact, fuel gas would not differ by much from the low conversion case.

The unit will be capable of a wide range of recycle cut points and recycle rates. This will allow selection of the best solvent for the reaction. Pilot work has been conducted to serve as a guide in solvent prepara-

tion and it is anticipated that further pilot work of this type may be performed. Design solvent cut points and solvent rate have been selected but it is expected that with variations in coal feed and product goals it will be useful to change these.

Yield pattern can be adjusted by changing the solvent cut points and rates. This is expected to be a useful tool in operation of the commerical unit. For instance, during periods of low feed rate the unit capacity could be used to convert heavy gas oil to lighter distillates.

The LC-Fining process is presently being operated commercially feeding petroleum residuum. There is little difference between the two operations in respect to reactor design and internals. Hydrogen consumption will tend to be a good deal higher when feeding solvent refined coal. This will in turn result in greater heats of reaction. The only effect on the first reactor operation will be that the mixed feed can be brought in at lower temperatures than is usual for petroleum operation. It will be necessary to use more quench material between the first and second reactor than is usual for petroleum operation.

The solvent refined coal will be de-ashed before feeding to LC-Fining. However, the LC-Fining reactor will be able to accommodate high ash levels with no problems within the reactor. This will be important during upset periods in the deashing unit. The pressure let down system will accommodate relatively short periods with high ash feedstock. Longer periods could result in some erosion of the let down valves and adjoining pipe, but it would take very long periods before maintenance was required. Such a long period would not be anticipated because the heavy product would need to be of low ash content for marketing which would dictate that any upsets in the deashing step be remedied relatively quickly.

Aging of catalyst with petroleum stocks occurs because of carbon deposition and the deposition of heavy metals such as vanadium and nickel which are contained in the feedstocks. Since there is no appreciable level of these metals in solvent refined coal, the aging of catalyst occurs due to carbon deposition.

Before feeding the solvent refined coal to the reactors, it is necessary to cut it back with a solvent. The solvent will be generated in the LC-Fining unit and separated out in the fractionation system for return to the suction of the feed pump. This operation is quite similar to commercial practice at Salamanca, Mexico where the gas oil is returned to the LC-Fining unit for conversion to light distillate.

Corrosion protection within the unit will be to the same standards as used when processing petroleum feeds. This includes appropriate metallurgy to handle hydrogen at high temperatures and pressure with additional protection for corrosion due to hydrogen sulfide, chlorides and polythionic acid. Reactors are of low chrome and molybdenum alloy with high chrome/nickel overlay. High temperature piping is of stabilized high chrome/nickel alloy.

Although complete details have not yet been worked out for the commercial unit to be installed at Newman, Kentucky, it is expected that it will be designed for feedstocks and conversions similar to those discussed here. Design and operation is expected to be similar to that employed for petroleum feedstocks in residuum hydrocrackers at Lake Charles, Louisiana and Salamanca, Mexico.

Table 1. LC-fining of solvent refined coal yields and quality 50% conversion case.

FEED	WT%	VOL%	API	SG	WT% IN FRACTION						LBS. PER 10,000 OF FEED						C/H
					S	N	O	H	C	ASH	S	N	O	H	C	ASH	
SRC IBP-850°F	6.81	7.60	-3.45	1.1050	.30	.96	2.00	7.80	88.94		2.00	6.5	13.6	53.1	605.7	—	
850-EBP	93.19	92.40	-17.70	1.2430	.73	2.13	4.63	5.77	86.59	.15	68.0	198.5	431.5	537.7	8669.3	14.0	
TOTAL	100.00	100.00	-16.70	1.2325	.70	2.05	4.45	5.90	86.75	.14	70.0	205.0	445.1	590.8	8675.0	14.0	
Products																	
H_2S	.61				94.08			5.92			57.1			3.6			
NH_3	1.06					82.24		17.76				87.1		18.8			
H_2O	3.77						88.81	11.19					335.0	42.2			
C_1	2.00							25.13	74.87					50.3	149.7		
C_2	2.00							20.11	79.89					40.2	159.8		
C_3	2.43							18.29	81.71					44.4	198.6		
C_4	2.29	4.83		.584				17.33	82.67					39.7	189.3		
C_5-390	10.60	15.46	36.0	.8448	.01	.40	1.35	10.83	87.41		.1	4.2	14.3	114.8	926.5		8.07
390-500	12.90	16.52	15.5	.9626	.04	.52	.60	9.60	89.24		.5	6.7	7.7	123.8	1151.2		9.30
500-650	7.80	9.26	4.8	1.0382	.06	.35	.70	8.46	90.43		.5	2.7	5.5	66.0	705.4		10.69
650-850	9.25	10.19	- 5.0	1.1186	.15	.75	.62	7.25	91.24		1.4	6.9	5.7	67.1	844.0		12.59
850-EBP	47.79	46.20	-20.5	1.2748	.22	2.04	1.61	4.81	91.03	.29	10.4	97.4	76.9	229.9	4350.5	14.0	18.93
TOTAL	102.50										70.0	205.0	445.1	840.8	8675.0	14.0	
C_4+	90.63	102.46	- 1.7	1.090	.14	1.30	1.21	7.08	90.11	.15	12.9	117.9	110.1	641.3	8166.9	14.0	
C_5+																	
650+	57.04	56.39	-18.0	1.2467	.21	1.83	1.45	5.20	91.06	.25	11.8	104.3	82.6	297.0	5194.5	14.0	

			RATIO OF CONCENTRATIONS FEED/PRODUCT			RATIO OF WEIGHTS FEED/PRODUCT				
Conversion, Vol% 850+	50.0									
Vol% 650+			S	N	O	S	N	O	Δ	API
Desulfurization, Wt%	81.6	C_4+								
Denitrogenation, Wt%	42.5	650+								
Deoxygenation, Wt%	75.3	850+								

$$H_2 \text{ Consumption, SCF/BBL} = \frac{H_2 \text{ Wt\% *Lb/Gal *42 *379.5}}{2.016 *100} = \frac{2.50 *10.267 *42 *3795}{201.6} = 2025$$

Table 2. LC-fining of solvent refined coal yields and quality 88% conversion.

FEED	WT%	VOL%	API	SG	WT% IN FRACTION					
					S	N	O	H	C	ASH
SRC IBP-850°F	6.81	7.6	-3.45	1.1050	.30	.96	2.00	7.80	88.94	
850-EBP	93.19	92.4	-17.7	1.2430	.73	2.13	4.63	5.77	86.59	.15
TOTAL	100.00	100.00	-16.7	1.2325	.70	2.05	4.45	5.91	86.75	.14

PRODUCTS

	WT%	VOL%	API	SG	S	N	O	H	C	ASH
H_2S	.72				94.08			5.92		
NH_3	2.13					82.24		17.76		
H_2O	4.45						88.81	11.19		
C_1	3.89							25.13	74.87	
C_2	3.89							20.11	79.89	
C_3	4.45							18.29	81.71	
C_4	4.73	9.98		.584				17.33	82.67	
C_5-390°F	26.20	38.22	36.0	.8448	–	.11	1.00	10.93	87.96	–
390-500°F	26.31	33.80	16.0	.9593	.01	.22	.35	9.65	89.77	
500-650°F	9.76	11.82	7.5	1.0180	.03	.25	.27	8.76	90.69	
650-850°F	8.40	9.51	- 1.5	1.0885	.05	.50	.26	7.72	91.47	
850-EBP	10.93	10.38	-22.5·	1.2982	.15	1.33	.91	5.00	91.33	1.28
TOTAL	105.86									
C_4+	86.33	113.71	19.7	.9357	.03	.35	.58	9.58	89.30	.16
650+	19.33	19.89	-13.4	1.1978	10	.97	.63	6.18	91.39	.73

Hydrogen Consumption, SCF/BBL	4750
Conversion, Vol. % 850+°F	88.3
Desulfurization, Wt. %	96.3
Denitrogenation, Wt. %	85.5
Deoxygenation, Wt. %	88.7

Table 3. Investment and operating requirements LC-fining of solvent refined coal.

Capacity, Ton/Stream Day	2289	17168
Investment, $MM	69	385
Utilities		
Cooling Water, GPM	3026	22700
High Pressure Steam, Net Consumed Lb/Hr	7100	53000
Medium Pressure Steam, Net Produced, Lb/Hr	24800	186000
Low Pressure Steam, Net Consumed, Lb/Hr	9200	69000
Electricity, KW	8900	67000
Fuel Gas, MMBTU/Hr	41.5	311.3

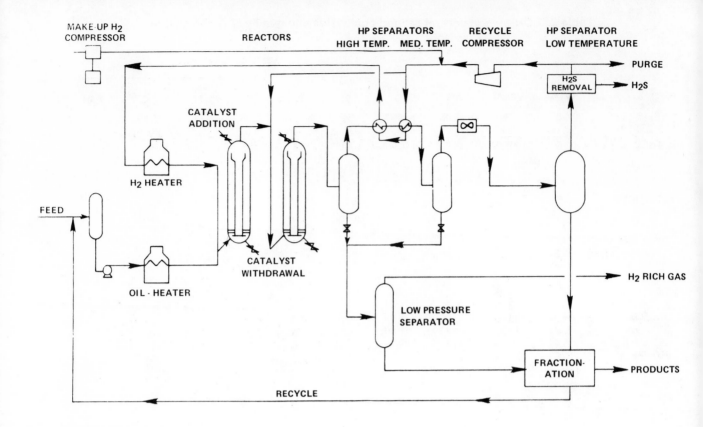

Figure 1. LC-fining of solvent refined coal.

CASPERS, J.

VAN DRIESEN, R.P.

HASTINGS, K.

MORRIS, S.

Hydrotreatment of COGAS Pyrolysis Oil via the H-Oil Process

The quality of raw oil generated from pyrolyzing coal is suitable for conversion into #2 and #6 fuel oil and a high-grade naphtha suitable for reforming to BTX or gasoline.

G. Nongbri and L. Lehman
Hydrocarbon Research, Inc.

L.I. Wisdom
COGAS Development Co.
Princeton, N.J.

The petroleum industry today has heavy demands on it to supply fuels for the motorists, for home heating, and for a variety of energy requirements for the economy, as well as to supply feedstocks for the petrochemical industry. The aromatics required by petrochemicals are being forced to compete with these heavy demands placed on the industry by the motorist. One source of easing this heavy demand and increasing the petrochemical supply picture is through the emerging synthetic fuel industry. Under a Department of Energy and Illinois Coal Gasification Group* (ICGG) contract, COGAS Development Company (CDC)** in conjunction with Hydrocarbon Research, Inc.*** have developed a process configuration for converting a pyrolysis oil from the COGAS Process into valuable fuel oils and a high grade naphtha suitable for reforming to BTX or gasoline.

The COGAS Process, developed by CDC for production of synthetic pipeline gas and liquids from coal, has been described in a number of papers and publications (1-10). A block flow diagram of the overall process concept is shown in Figure 1. A COGAS Process Demonstration Plant is currently being designed under a contract between ICGG and the Department of Energy.

Briefly, the process steps are as follows: The washed coal is prepared for processing by being crushed, screened and pressurized. The full size consist of coal is fed to fluidized bed pyrolyzers where it is dried and heated in stages to drive off volatile matter. The residue, or char, flows to the gasifier where it is gasified in a fluidized bed with steam. The product gasifier gas provides sensible heat and fluidizing gas for pyrolysis. The gas and oil vapors from pyrolysis are sent to oil recovery where they are quenched and the condensed oil is separated from the synthesis gas. The oil is hydrotreated to produce naphtha and fuel oil. The raw synthesis gas is purified to remove hydrogen sulfide and carbonyl sulfide, and then the gas undergoes shift and methana-

---

*A partnership of subsidiaries of Northern Illinois Gas Company, The Peoples Gas Light and Coke Company, Central Illinois Public Service Company, Central Illinois Light Company, and North Shore Gas Company.

**A partnership of Consolidated Gas Supply Corporation, a subsidiary of Consolidated Natural Gas Company, FMC Corporation, Panhandle Eastern Pipe Line Company, and Tennessee Gas Pipeline Company, a Division of Tenneco Incorporated.

***A subsidiary of Dynalectron Corporation.

tion, bulk carbon dioxide removal, dehydration and compression to a pipeline quality gas. A portion of the purified synthesis gas is used to provide hydrogen required for hydrotreating the raw oil. Each of the process steps is designated by a plant area number (e.g. Area 104B, Hydrotreating).

The first phase of the DOE-ICGG Demonstration Plant Program is subdivided into 11 specific tasks. The first of these, Task I, is entitled Conceptual Design and Evaluation of a Commercial Plant.

The Task I work scope was divided into four parts:

1. Tentative Baseline Design Report
2. Process Evaluation Report
3. Revised Tentative Baseline Design
4. Conceptual Commercial Plant Design Report

All four of these milestones have been completed.

The Tentative Baseline Design (TBD) was submitted to DOE in August 1977. It included process flow diagrams, process descriptions and mass/energy balances for a commercial plant concept. This "baseline" concept was then expanded to a more complete design, including equipment specifications, utility summaries, steam-power balances, layout drawings and cost estimates. Thus, a complete technical and economic baseline was developed in which process optimization and trade-off studies could be made.

The design of Area 104B in the TBD, Hydrotreating, was based on COED* hydrotreating pilot plant tests which processed 4,000 barrels of filtered pyrolysis oil from eight different coals in a 30 bbl/d fixed bed reactor. Because of problems in filtering a pyrolysis oil and because of the low catalyst life resulting from the deposition of soluble metals and carbon on the catalyst, two ebullated-bed reactors in series followed by a fixed-bed reactor were chosen for the TBD. The first ebullated-bed reactor acted as a guard chamber where metals were removed, and the second reactor was the main

_____
* The COED Process was developed by FMC Corporation under an OCR and ERDA contract between 1962 and 1975 (15).

hydrotreater. The solids in the feed oil were removed after the ebullated-bed reactors by vacuum filtration.

To confirm the design criteria of an ebullating-bed reactor when processing a pyrolysis coal-derived oil, Hydrocarbon Research, Inc. was contracted by CDC to carry out single-stage bench-scale tests in order to develop design criteria for the hydrogenation of unfiltered pyrolysis oil in an H-Oil reactor.

In conjunction with this effort by HRI, CDC conducted a trade-off study which considered the production of fuel oil and naphtha versus syncrude and vacuum filtration versus vacuum distillation for removing the solids (the base case used in the TBD), Figure 2. Results from this study overwhelmingly favored the production of fuel oil and naphtha, and the use of a vacuum distillation tower for removing the solids from the oil.

Following the successful H-Oil bench-scale hydrotreatment of pyrolysis coal-derived oil in a single stage unit, HRI was contracted for the conceptual design of a commercial and demonstration plant that would hydrotreat an Illinois, a Pittsburgh seam, and a Western subbituminous pyrolysis coal-derived oil to fuel oil and naphtha. The results from both of these studies are presented herein.

H-OIL PROCESS

The H-Oil Process is a proprietary commercial catalytic hydrogenation process developed by Hydrocarbon Research, Inc. The original application envisioned for the H-Oil Process was the upgrading of hydrogen-deficient petroleum residua, but the flexibility of the process made it possible to extend its application to liquefy coal (H-Coal Process) and to upgrade coal liquids.

Since the H-Oil Process and its application in a refinery have been described in detail elsewhere(11, 12) only a brief description is presented. The heart of the H-Oil Process is the ebullated-bed reactor. Figure 3 illustrates the basic features of the H-Oil reactor. The process features an ebullating bed of catalyst in which the upward velocity of the gas and liquid maintains the expanded bed of catalyst in continuous random motion. The main advantages of this system are:

1. Extraneous inert solids introduced with the liquid feed are passed through the system and discharged with the product.

2. Removal of inorganic matter by filtration, as required at times in a fixed-bed system, is not necessary.

3. Isothermal conditions are maintained throughout the reaction space in the presence of highly exothermic reactions.

4. Catalyst can be added and withdrawn during operation. This prevents down time for charging the catalyst and produces constant quality product because of constant catalyst activity throughout the duration of operation.

Solids and any other material that plug fixed bed reactors end up in the heavy ends of the product and are removed from the system along with the vacuum bottoms product which is sent to a combustor.

CHARACTERISTICS OF RAW PYROLYSIS OIL

The unfiltered pyrolysis oil produced from the COED Process had a high specific gravity (-4 to -6 °API), a low hydrogen content and high concentrations of oxygen, nitrogen, and sulfur. The char fines entrained in the raw oil range from 2 to 10 weight percent (measured as quinoline insolubles) and average about 5 microns in size.(13) Detailed inspections of the raw oil from Western Kentucky coal are shown in Table 1.

The raw pyrolysis oil cannot be used as a fuel oil or petrochemical feedstock because of its properties and high concentration of contaminants. Hence, hydrotreating of this product over a catalyst is required in order to improve its properties and marketability. During hydrogenation, hydrogen is added to the oil, and nitrogen, oxygen, and sulfur contents of the oil are reduced to environmentally acceptable levels. Furthermore, the

viscosity is reduced, and the storage stability of the oil is improved to a satisfactory level.

BENCH SCALE OPERATION

The objectives of the H-Oil bench scale program were to:

a) Establish the feasibility of processing raw unfiltered pyrolytic oil in an ebullated-bed reactor,

b) Determine the operating conditions necessary for hydrogenation,

c) Obtain data on hydrogen consumption, yields, distribution of products, catalyst life and degree of heteroatom removal.

The work was performed in a single-stage continuous H-Oil bench-scale unit over a sample of commercial H-Oil catalyst. A simplified diagram of this unit is shown in Figure 4. This unit simulates the operation of a full-scale H-Oil unit in all important aspects, and is used routinely for H-Oil catalyst testing and feedstock studies.

OPERATING HISTORY

The operation of the bench scale unit was highly successful. Throughout the duration of the run, the operation was smooth with no operating problems. As expected, the solids content in the feed did not create any problems either in the preheater or inside the catalyst bed. Following shutdown, the catalyst discharged freely from the reactor and was in free-flowing condition and no catalyst agglomeration was found. The reactor and product lines were clean, indicating that the conditions selected along with the feed/catalyst combination were ideal for this system.

EXPERIMENTAL RESULTS

This program processed an unfiltered pyrolysis oil from a Western Kentucky bituminous coal. The operating conditions were selected to maximize heteroatom removal and to minimize the production of light hydrocarbons and 900°F+ (482°C+) fraction of the oil. The severity of H-Oil operations is measured by the amount of the 900°F+ (482°C+) fraction that cracks into a lighter material. The conversions studied in this operation ranged from 72 to 93 liquid volume percent.

At the above conversion levels, sulfur removal was better than 90 percent whereas oxygen removal ranged from 72 to 85 percent. Nitrogen removal ranged from 33 to 69 percent.

Because of the high concentration of iron (believed to be organometallic type) in this sample of oil, it was feared that the catalyst would show a high rate of deactivation. Results from the test did not confirm this assumption. Catalyst deactivation measured in terms of nitrogen removal as function of catalyst age (defined as barrel of oil per lb of catalyst), is shown in Figure 5.

The yields of gas and liquid products from this operation are summarized in Figure 6 as function of 900°F+ (482°C+) conversion. Detailed product yields and properties from an operation at 80 percent 900°F+ (482°C+) conversion are summarized in Table 2.

Chemical hydrogen consumption ranged from 2100 to 2780 scf/bbl, 374 to 495 cu m of hydrogen per cu m of oil. A summary of average hydrogen consumption as function of 900°F+ (482°C+) conversion is shown in Figure 7.

At the end of the bench scale experimental program, HRI prepared a turbine fuel sample by distilling the product oil into the 350-900°F (177-482°C) boiling point range. Properties of this material are shown in Table 3. The combustion characteristics of this sample were measured in a mini combustor by a turbine manufacturer under an industry sponsored program.

The results from these tests indicated that a smooth transition occurred from the baseline petroleum fuel (No. 2 oil) to the sample product from the H-Oil reactor. Combustion exit temperatures remained constant while carbon monoxide, hydrocarbons, and smoke emissions remained at very low and unchanged levels during the combustion tests. As expected, the NOx levels at the various combustion temperatures tested were higher than the NOx levels produced from

petroleum No. 2 oil. The lighter fraction of the raw oil in a COGAS plant will undergo two stages of hydrotreatment (H-Oil reactor followed by a fixed bed reactor). As a result, the nitrogen content in the turbine fuel portion of the product will be reduced to levels that could match NOx levels produced from combusting a petroleum No. 2 oil. Alternatively, combustion controls can be added to minimize NOx emissions.

CONCEPTUAL COMMERCIAL PLANT DESIGN:

Hydrotreating Area

A simplified flow scheme of the H-Oil Process for hydrotreating pyrolysis oil for a COGAS plant is shown in Figure 8. Hydrogen rich gas which has been heated to the appropriate temperature is combined with hot unfiltered pyrolysis oil and fed to the H-Oil Reactor. The mixture is mixed with circulated reactor liquid in a plenum chamber at the bottom of the reactor. This mixture passes through a specially designed distributor plate which insures proper distribution of vapor and liquid over the reactor cross-sectional area. Above the distributor the mixture passes upward through a bed of H-Oil catalyst. The action of the upward flowing liquid maintains the catalyst in a fluidized state referred to as an ebullated bed. The reaction occurs in the liquid phase and the exothermic reactions provide sufficient heat to maintain the reactor at the desired temperature level.

The expansion of the catalyst bed is controlled by the circulation of reactor liquid. This circulation is normally several times the rate of fresh feed liquid. A specially designed ebullating pump is installed in the bottom of the reactor, either internally or externally but contiguous with the reactor. This pump is usually of a "canned" design with a variable speed electric motor to control the rate of flow. It is this ebullating pump that gives H-Oil a feature which is unique among fluidized systems. By varying the internal ebullating flow rate, proper expansion of the catalyst bed can be maintained over a wide range of raw oil feed rates to the reactor as well as over a wide range of operating conditions.

As shown in Figure 3, the ebullating

pump takes suction from a recycle tube in the center of the H-Oil reactor. As part of the H-Coal program, a major study was made of the hydraulics of this system. Present designs of the downcomer pan at the top of the recycle tube are substantially different from that in present day commercial H-Oil operations. The new downcomer pans permit operation at higher vapor throughputs than the original design.

The inspection of the products from H-Oil hydrotreating given in Table 4 show that the distillate fractions are low in sulfur but high in nitrogen. This naphtha can be reformed either to a high-octane gasoline or to BTX. In a reforming process, the necessary reactions must take place over platinum-containing catalyst. The catalysts are generally poisoned by nitrogen compounds. Generally, catalytic reforming units have provision for pretreatment of the feedstock to remove sulfur compounds. In many refineries these pretreaters are capable of substantial nitrogen removal. However, the level of nitrogen encountered in petroleum refining is far less than the nitrogen level shown above for coal-derived naphtha from the H-Oil Process. In addition, present-day requirements are limiting the amount of nitrogen that can be left in fuel oils.

The H-Oil reactor effluent is largely in the vapor phase and contains hydrogen at high pressure as well as most of the naphtha produced in the H-Oil Process, along with some of the distillate fuel oil compounds. It is relatively inexpensive to provide for a fixed bed of an appropriate catalyst, such as catalyst containing nickel and molybdenum, for the purpose of nitrogen removal. Therefore, a vapor-phase fixed-bed reactor was added inline with the vapor stream in order to remove the nitrogen in the naphtha to less than 1 ppm.

Effluent from the H-Oil reactor is separated into a vapor and liquid phase, with the vapor phase then being further cooled to a lower temperature for feed to the fixed-bed hydrotreating reactor. The condensate formed in this cooling is removed from the feed to the fixed-bed hydrotreating reactor for the purpose of avoiding unnecessary hydrogen consumption and catalyst deactivation.

The condensate is sent to a vapor/liquid separator and then to a fractionator where the dissolved naphtha is removed and sent to the fixed-bed hydrotreater. The heavy oil fraction from the fractionator is sent to a vacuum tower where the 900°F+ (482°C) fraction containing the char solids is removed and sent to a combustor. The remaining oil is fractionated into a recycle seal oil and product fuel oil.

The vapor from the fixed-bed hydrotreater reactor is then cooled by a series of exchangers, and finally by air or water to ambient temperature. Process water must be injected into this stream before final cooling to avoid precipitation of ammonium salts. Three phases are separated in the flash drum. The sour-water phase contains ammonium salts, and this sour water goes to a system that will produce saleable ammonia along with other by-products from the COGAS Process. The vapor phase from this separation is largely recycled hydrogen which is returned to the reactor. A small portion of this impure hydrogen is purged to avoid buildup of gaseous impurities.

The third phase from this flash drum is the hydrotreated liquid hydrocarbons containing most of the naphtha produced in the H-Oil Process, as well as some of the distillate fuel oil. This liquid is fractionated into naphtha and a distillate fuel oil. The distillate fuel oil is combined with the other fuel oil fractions and sent to product storage. Estimated compositions of the liquid products are shown in Table 5.

ECONOMICS

With the substitution of HRI's commercial hydrotreating design and with the change in the product slate from syncrude to fuel oil and naphtha, the capital cost for this area was reduced by 45 percent. This capital cost translates into a unit cost of $3.69/bbl ($23.21/cu m) for fuel oil and naphtha. Of course, there are other capital-related costs involved with the production of liquid fuels, but many of these are interrelated with the production of synthetic pipeline gas.

Several other areas besides hydrotreating were improved as a result of the process and trade-off studies. All of these improvements were included in the Conceptual Commercial Design Report issued in September 1978.

In the Conceptual Commercial Plant, a self-contained grass roots facility was designed to produce 264.5 million cu ft/d (7.49 million cu m/d) of synthetic pipeline gas from 25,935 TPD (23.5 gigagrams) of Illinois No. 6-seam coal, raw water and miscellaneous catalyst and chemicals which constitute the total feed to the plant. All utilities, including electric power, are produced onsite, except the electric power needed to operate the river water pumping station and miscellaneous plant lighting. By-products from the plant are in finished marketable form and include the following daily production:

o   16,823 barrels (2675 cu m) of fuel oil
o   3,915 barrels (622 cu m) of naphtha
o   2,178 tons (1,976 megagrams) of sulfuric acid
o   48 tons (43.5 megagrams) of ammonia

The capital requirements and operating costs for the conceptual plant design are based on a site in Perry County, Illinois. This provides the basis for fixing various economic factors such as the cost of coal, cost of labor, labor productivity, transportation, etc.

The commercial plant capital requirement is estimated at $1,643 million (mid-1978 dollars). The plant has a total daily output of 378.7 billion Btu (399.5 terajoules), of which fuel oil and naphtha represent approximately one third of the total. These costs are shown in Table 6. The total plant investment cost (which is the capital requirement less working capital, initial charge of chemicals and catalysts, cost of land, start-up, royalties and administration) amounts to $1,199 million, or 80.9 percent of the total capital requirement.

The operating cost for the commercial plant is based on a 20-year plant design life, 330 days per year of operation, with an overall thermal efficiency of 65.3 percent. The operating cost is divided by classifications as shown in Table 7 for a total cost of $281 million per year.

There are many bases on which one may calculate gas costs. One typical set for a utility company is shown in Table 8. Using these guidelines with the capital and operating costs for the Conceptual Commercial Plant, a first year pipeline gas selling price of $5.30 per million Btu ($5.03 per gigajoule) was calculated, Table 9. Since September 1978, fuel oil and naphtha have increased from $15.40/bbl ($96.9/cu m) and $16.80/bbl ($105.7/cu m) to $29.12/bbl ($183.3/cu m) and $33.56/bbl ($211.1/cu m), respectively, as of January 16, 1980 in Chicago, Illinois.(14) This change in price reflects a 53 percent annual growth rate. In the COGAS process, the revenues received from fuel oil and naphtha are credited towards the total required revenues from the plant. As shown in Figure 9, a 53 percent increase in fuel oil prices would reduce the gas price by 23 percent to $4.12 per million Btu ($3.91 per gigajoule). Of course the magnitude of the decrease in gas price would partially be offset by the increases in capital and operating costs over the last two years. In essence, the rapid rise in crude oil prices and subsequent increases in fuel oil prices have acted as an inflation hedge by keeping the price of synthetic natural gas down for the COGAS Process.

ACKNOWLEDGMENT

The authors wish to thank Mr. Jeffrey L. Gendler of HRI for his contribution to the H-Oil bench scale experimental work.

LITERATURE CITED

1. Friedman, L. D., "Development of a Fluidized Bench-Scale Reactor for Kinetic Studies," presented at American Chemial Society Meeting, Division of Fuel Chemistry, Chicago, Illinois, (1975).

2. Eddinger, R. T., "Pyrolysis Route to Coal Conversion," presented at the World Coal Conference, London, England, (1975).

3. Hoy, H.R., Eddinger, R.T., "COGAS Progress Report, 1975," presented at the AGA Seventh Synthetic Pipeline Gas Symposium, Chicago, Illinois, (1975).

4. Bloom, R. Jr., "The Illinois Coal Gasification Group Project Incorporating the COGAS Process," presented at the

American Gas Association Eighth Synthetic Pipeline Gas Symposium, Chicago, Illinois, (1976).

5. McCray, R. L., McClintock, N., Bloom, R. Jr., "The Illinois Coal Gasification Group, What is it and Where is it Going," presented at the American Institute of Chemical Engineers, Philadelphia, Pa., (1978).

6. Eddinger, R. T., Bloom, R.Jr. "The COGAS Process--A Promise for the Future," University of Pittsburgh Conference, (1978).

7. Eby, R. J., McClintock, N., Bloom, R. Jr., "The Illinois Coal Gasification Group Project - COGAS Process," presented at the Tenth Synthetic Pipeline Gas Symposium, Chicago, Illinois, (1978).

8. Bloom, R. Jr., "Coal Dilemma II," presented at the American Chemical Society, (1979).

9. Bloom, R. Jr., and Wisdom, L. I., "Chemical Feedstocks From Coal," presented at the AIChE, Houston, Texas, (1979).

10. Wisdom, L.I., and Ketkar, A.B., "The COGAS Process - Clean Energy From Coal," presented at the Sixth National Conference on Energy and the Environment, (1979).

11. Chervenak, M.C., Johnson, C.A. and Schuman, S.C., "H-Oil Process Treats Wide Range of Oils," Petroleum Refiner, 39, (10), p. 151-6, (1960).

12. Brawley, J. M., "H-Oil Units Attain High Throughput, Long Runs," Oil and Gas Journal, (August 19, 1974), p. 65.

13. Merrill, R.C., Scotti, L.J., Ford, L. and Domina, D.J., "The Production of Clean Fuels from Eastern Coals by the COED Process," presented at the 79th National AIChE Meeting, Houston, (1975).

14. Shumway, D. L., "The Oil Daily," (Jan. 16, 1980).

15. "Char Oil Energy Development," Final Report, September 1975, U.S. ERDA Contract No. E(49-18)-1212, FE/1212/F, District Category UC-90D.

NONGBRI, G.

LEHMAN, L.

WISDOM, L.I.

Table 1. Properties of unfiltered coed pyrolysis oil.

Coal Source	West Kentucky
Gravity, °API	-6.2
Carbon, W %	81.70
Hydrogen, W %	7.67
Sulfur, W %	1.61
Nitrogen, W %	1.10
Oxygen, W %	7.49
Ash, W %	0.43
Iron in Ash, W %	20.1
Quinoline insolubles	2.0
Viscosity, SUS @ 210°F	41
Metals, Wppm	
Iron	757
Vanadium	Nil
Nickel	Nil
Distillation (ASTM D1160)	
IBP, °F	408
V % @ 400°F	0
V % @ 650°F	24
V % @ 900°F	59.7
V % @ 975°F	69.3

Table 2. H-Oil® processing of coed pyrolysis oil from west Kentucky coal H-Oil bench scale reactor data.

FEED PROPERTIES

Gravity, °API	-6.2
Sulfur, W %	1.61
Nitrogen, W %	1.10
Oxygen, W %	7.49
Ash, W %	.43
Quinoline Insolubles, W %	2.0
900°F + Fraction, LV %	40.3

OPERATING DATA

900°F + Conversion, LV	81.3
Desulfurization, W %	98
Denitrogenation, W %	53
Chemical H_2 Consumption, SCF/bbl	2610

YIELDS AND PRODUCT PROPERTIES

	W %	V %	°API	Wt% S	Wt% N	W% C	W% H	Viscosity SUS @ 122°F	CPS@300°F	W%RCR	Br.No.	Pour Point °F
H_2S	1.7											
NH_3	0.7											
$CO + CO_2$	0.7											
H_2O	6.1											
C_1	1.4											
C_2	1.5											
C_3	1.8											
C_4- 400°F	16.2	24.3	54.2	-	0.147	86.60*	12.71*				14.7*	
400-650°F	39.2	46.6	17.4	-	0.523	87.00	10.49	38			50.3	-25
650-900°F	26.7	29.4	6.4	< 0.03	0.787	87.89	8.72	355				75
900°F+	7.6	7.5	-7.7	0.18	0.956	88.53	7.70		59	22.3		
TOTAL	103.6	107.8	19.1	0.03	0.57							

* 175-400 Fraction

Table 3. Properties of turbine fuel from H-Oil bench scale operation on unfiltered pyrolysis oil from a western Kentucky coal.

Nominal Boiling range, °F		350-900
Gravity °API		12
Sulfur	W%	< 0.02
Nitrogen	W%	0.6*
Carbon	W%	87.9*
Hydrogen	W%	10.0*
Distillation D-1160		
IBP°F		378
10V%		450
30V%		537
50V%		609
70V%		708
90V%		826
EP		968
V% @ 900°F		96

* Estimated Values

Table 4. H-Oil® processing of COGAS pyrolysis oil from Illinois #6 coal estimated yields from an H-Oil reactor.

OPERATING DATA

900°F + Conversion, LV %	77
Desulfurization, W %	96
Denitrogenation, W %	46
Chemical Hydrogen Consumption, SCF/bbl	2500
Catalyst Replacement Cost, ¢/bbl (1980 Net $)	36

YIELDS AND PRODUCT PROPERTIES

Fractions	W %	V %	°API	W% S	Wppm N
H_2S	2.2				
NH_3	0.6				
$CO + CO_2$	0.6				
H_2O	7.1				
C_1	1.1				
C_2	1.2				
C_3	1.3				
C_4- 400°F	15.0	21.8	54	-	0.15
400-650°F	39.1	45.9	18	-	0.55
650-900°F	25.1	27.3	7	< 0.03	0.79
900°F	10.1	9.5	-12	0.65	1.03
TOTAL	103.4	104.5	17.7	0.08	0.60

Ash in C_4 + Liquid = 0.50 W %

Table 5. Estimated chemical and physical properties of product fuel oil and naphtha from an Illinois #6 pyrolytic coal oil.

	No. 4 Fuel Oil	Naphtha
Composition, wt %		
Carbon	88.02	86.62
Hydrogen	10.65	13.38
Nitrogen	0.36	0.00
Sulfur	0.03	0.00
Oxygen	0.94	0.00
Total	100.00	100.00
Molecular weight	266.0	93.6
Pour Point	-30	-
Flash point, °F	175	-40
Vapor Pressure, mm Hg		
@ 80°F	.65	315[1]
@ 120°F	1.39	658
Distillation, °F		
0 - vol %	244[2]	100.3[3]
5	365	106.5
10	420	120.7
30	515	175.8
50	600	216.5
70	675	260.3
90	800	331.2
95	875	361.9
100	1000	401.7
°API	13.89	49.09
UOP Characterization Factor, K	10.51	11.15
Viscosity, 100°F		
SUS	47	
CS	6.4	

[1] High vapor pressure due to large quantities of butane in naphtha stream.

[2] True boiling point

[3] ASTM

Table 6. COGAS process commercial plant capital requirement.

Capital Requirement	Costs (Thousands of Dollars)	Percentage
Plant Investment	$1,199,310	73.0
Contingency, 15%	177,368	10.8
Construction Loan Interest	170,901	10.4
Land	2,039	0.1
Start-up	26,694	1.6
Administration	4,150	0.3
Working Capital	30,857	1.9
Royalties	20,000	1.2
Initial Charge of Chemicals and Catalysts	12,154	0.7
Total Capital Requirement	$1,643,473	100.0

Mid-1978 Dollars

Table 7. COGAS process annual operating cost for a commercial plant.

Classification	Annual Cost (Thousand of Dollars)
Coal @ $22.85/ton	$195,563
Labor	23,138
Chemicals and Catalysts	20,769
Insurance and Taxes	7,837
Repairs and Replacements	30,427
Utilities	1,060
Other Operating Supplies	2,293
Gross Operating Costs	$281,086

Table 8. Basis for first year economic analyses utility-type financing.

Total capital includes	- Fixed capital
	- Construction loan interest
	- Land
	- Start-up costs
	- Administration
	- Working Capital
	- Initial charge of catalysts and chemicals
Coal Price	- Illinois, $22,85/ton
Labor	- $9.00/hr
Debt/Equity ratio	- 75/25
Interest on debt	- 9%
Depreciation	- 5% per year, straight line
Return	- 15% on equity, after tax
Federal income tax	- 48%
Sulfuric Acid Credit	- $57.3/metric ton ($52/ton)

Table 9. Gas price calculations utility type financing-mid 1978 dollars Illinois #6 coal.

Total Capital Requirement	$1,643.5 MM
Coal: 25,935 ton/day	195.56
Operating Cost	85.52
Gross Operating Cost	$281.08
Depreciation (5%/yr)	80.63
Interest: 75% debt @ 9%	108.21
After-tax profit: 25% equity @ 15%	60.12
Federal Income tax (48%)	55.49
Total Required Revenue	$585.53

Products	$/(unit)	Total ($MM/yr)
No. 4 Fuel Oil - 16,823 bbl/d	15.40	85.49
Naphtha - 3,815 bbl/d	16.80	21.15
Sulfuric Acid - 2,178 tons/d	52	37.37
Ammonia - 47.9 tons/d	120	1.90
		145.91
First Year SPG Price - 251.4×10^9 Btu/d	5.30	439.62
20 Year Avg. Gas Price	4.01	
First Year Avg.-Energy Products Price	4.37	
20 Year Avg. Energy Products Price	3.52	

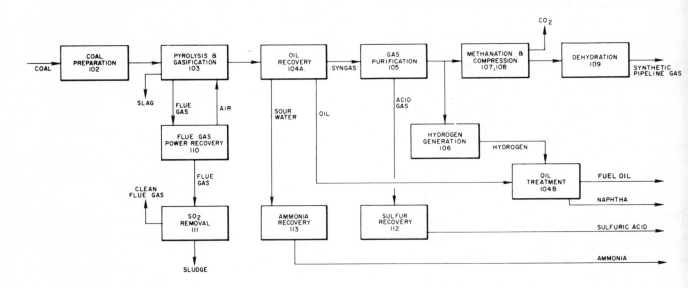

Figure 1. COGAS process block flow diagram.

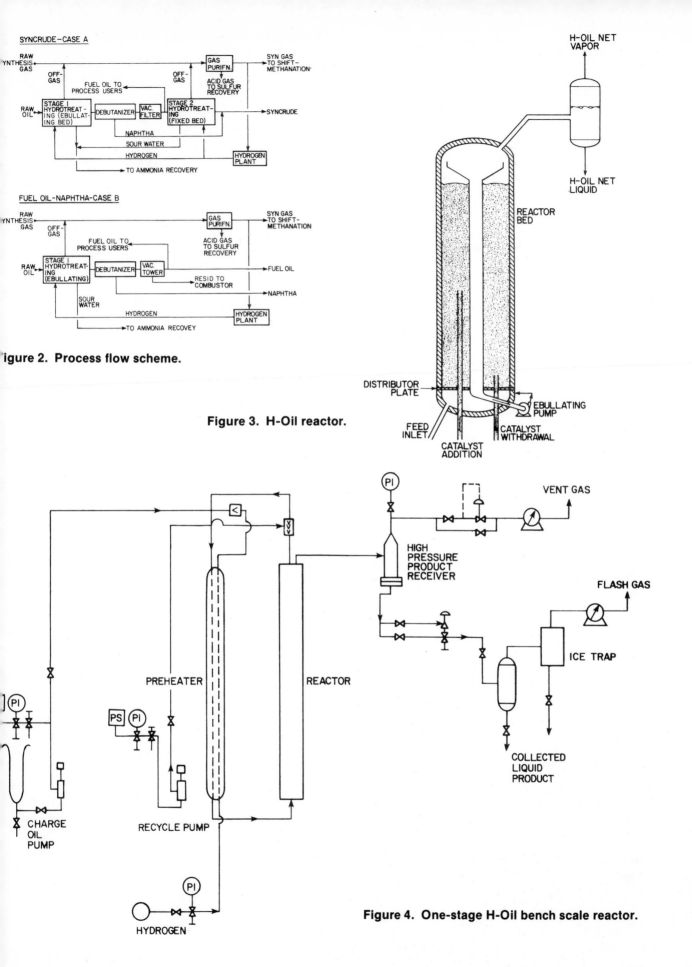

SYNCRUDE—CASE A

Figure 2. Process flow scheme.

Figure 3. H-Oil reactor.

Figure 4. One-stage H-Oil bench scale reactor.

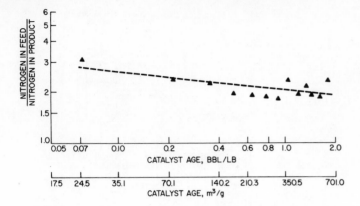

Figure 5. H-Oil processing of unfiltered pyrolysis oil: nitrogen removal as a function of catalyst age.

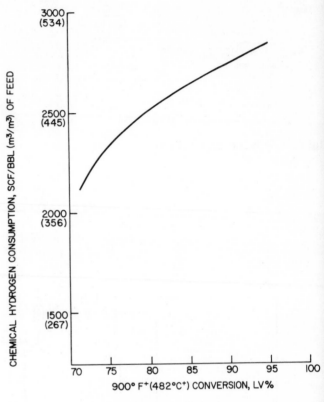

Figure 7. H-Oil processing of unfiltered pyrolysis oil: variation of chemical hydrogen consumption with 900°F⁺ conversion.

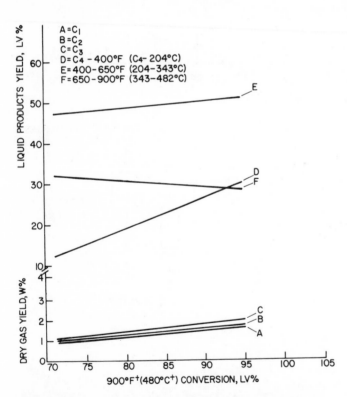

Figure 6. H-Oil processing of unfiltered pyrolysis oil: dry gas and distillate yields. 900°F⁺ conversion.

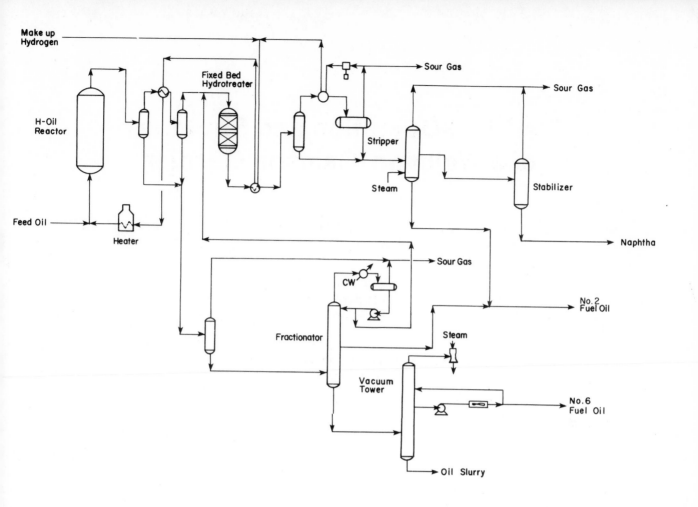

Figure 8. Process flow diagram: hydrotreating.

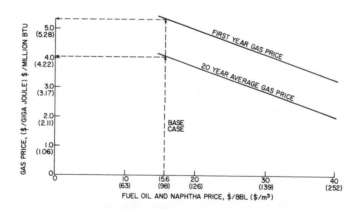

Figure 9. SPG gas price as a function of fuel oil and naphtha price utility type financing.

Start-Up, Operation and Shutdown
of a Wellman-Galusha Gasifier

Operating experience has shown this gasifier to be an extremely stable piece of equipment.

A.C. Mengon
Dravo Corp.
Pittsburgh, Pa.

J.S. Levin
Fluor Engineering and Constructors
Houston, Tx.

This discussion is intended as an introduction to the operability of a small air blown coal gasifier. The gasifier discussed is a Wellman-Galusha , as offered by the Dravo Corporation, Pittsburgh, Pennsylvania. It is 6.5 feet (1.98m) internal diameter. It is installed at the U.S. Bureau of Mines experimental station in Minneapolis, Minnesota, to supply hot, raw gas to an iron ore pellet induration kiln. A brief description of the gasifier and its auxiliaries are given here. They will be followed by a description of the operating procedures used during start-up, steady state operation, load variation, shutdown and clean-out. A review of key operating parameters is presented. Control of the gasification system is made possible through interpretation of this key information.

DESCRIPTION OF MACHINE

The Wellman-Galusha is a fixed bed, single stage gasifier. "Fixed bed" does not mean that the fuel bed is stationary. On the contrary, coal is continuously fed at the top, and ashes are discharged continuously from the gasifier to the integral ash pit, from where they are withdrawn as needed. "Fixed bed" refers to the gasification conditions. These remain "fixed" for long periods of time and are changed only as the fuel changes and/or the gas demand changes.

There are two types of Wellman-Galusha gasifiers: those equipped with an internal agitator, and those without an agitator. The agitator rotates and it also reciprocates in the vertical direction. This enables an agitator equipped gasifier to handle caking coals. An additional benefit is an increase in capacity, in some cases up to about 25% more than a non-agitated gasifier.

The Wellman-Galusha installed at Minneapolis is an agitated gasifier. It was originally installed at Hanna Mining Company's Hibbing, Minnesota, ore plant as a research tool, much in the manner of its current use, i.e., to provide gas from various test fuels, the gas to be used in induration tests of iron ore pellets. Figure 1 gives construction details of the Wellman-Galusha gasifier. Figure 2 is a schematic of the pelletizing pilot plant.

The gasifier body is water jacketed around the sides and over the top. Note that a space exists between the top of the water and the top of the gasifier. Passage of the air blast over this water provides the steam necessary for control of the air blast temperature. The selection of and the control of the proper "saturation" temperature, as the temperature of the mixed steam-air stream is called, is critical in the avoidance of clinkering problems.

The grate, located in the bottom of the gasifier body, is a three tiered assembly consisting of top "cone", and second step "washer" mounted off-center and a concentric lower "washer" mounted on center. It is shown on Figure 3. Rotation of the unit creates a "slicing" action where the cone and the middle plate "slice" through the ash. The eccentric motion causes the ash cut out of the ash bed to fall into the ash pit from where it is removed.

An air blower is provided to supply the air blast needed for gasification.

"Poke holes" are access hatches welded into the top head of the gasifier. They permit visual inspection of the surface of the bed. Their main purpose is to allow the insertion of steel rods approximately 3/4" in diameter into the bed. Figure 4 shows the main features of the poke hole. Figure 5 depicts a fire test rod. Inspection of these rods and the interpretation of the data gained therefrom are the main pieces of information used to control the gasifier.

The lock hoppers shown in Figure 6 are equipped with large disc valves that permit rapid fill-up and at the same time are synchronized to preclude direct discharge of gas during the charging sequence. The size of the feed pipes serves to maintain a coal feed, even during charging of the hoppers.

The cyclone removes entrained solids from the off-gas. It is constructed so that filling its lower portion with water creates an effective seal when it is desired to isolate the gasifier.

An off-gas vent valve and flare are provided through which waste gases are vented and burned.

OPERATING PROCEDURES

Start-Up

Preparation of the gasifier for start-up begins with laying down a 12-18" (30-45cm) ash layer over the grate. This serves as an initial ash bed for protecting the grates from the hot fire zone. "Fire boxes" constructed beforehand consisting of (6" sq x 18" lg) (15 cm x 46 cm) wooden boxes stuffed with burlap are placed on the bottom level of the grate at equal intervals around the gasifier perimeter and around the middle tier of the grate. Each box trails a tail of burlap which hangs below the grate and into the ash pit

just above the ash pit valve. These tails are tied together into a single tail which acts as the fuse. A 1/2"(14mm) diameter pipe is run through the poke holes to the top of the fire boxes. These pipes are used to add fuel oil into the boxes immediately prior to light off. Above the ash, an 8"-10" (20-25 cm) thick layer of charcoal is added. After the charcoal is in place, it is sprayed with fuel oil. The gasifier is buttoned up and the coal hoppers and feed system are filled with coke and operated to fill the gasifier. The water jacket is filled; a minimal overflow is maintained. Fuel oil is added to the fire boxes via the pipes and is sprayed onto the fuse via the fire doors on the side of the gasifier at grate level. To prepare the remainder of the system for light off, the vent valve is opened and the flare is ignited. Also, the cyclone is filled to the overflow level with water and the grate and the grate and agitator are set at minimum speeds. To light off, the fuse below the grate is lit and the fire door is closed. Simultaneously, the air blower is turned "on" at a minimum rate. Figure 7 shows the arrangement of the gasifier ready for light-off.

With the gasifier lit and running on coke and with off-gas being vented, the bed is monitored for stability for a reasonable length of time (4-8 hours) before switching the gas flow from the vent to the end user. Conditions to be checked before switching are 1) gas composition , 2) saturation blast temperature, 3) bed level, 4) off-gas temperature, and 5) ash level. Once stable operation is achieved, a switch-over to the user is performed. This switchover is accomplished by allowing the level of water in the cyclone to fall (drain) while simultaneously "pinching off" the vent valve. Care must be taken so as not to shut the vent valve too quickly, blowing the water seal on the cyclone. These operations are coordinated by monitoring the pressure immediately upstream from the cyclone. Once the vent valve is closed and the water level in the cyclone is at the required seal height, the system is in operation.

Steady State Operation

Operation of the producer under steady state conditions requires monitoring several key parameters as discussed below. It is important to note that due to the high capacitance of the producer system, sudden shocks to the gasifier system are dampened advantageously.

To maintain the producer in a stable operating condition requires monitoring the key parameters listed below. The instrumentation used to control the gasifier is shown on Figure 8.

1. Pressure drop across the ash and coal bed: This is monitored by appropriate pressure guages located on the ash pit and on the exit gas off-take line.

2. Off Gas Temperature: This is monitored by thermocouples located upstream of the cyclone separator on the exit gas off-take line.

3. Ash Bed Level: This is measured by inserting steel rods into the producer for a period of 3-5 minutes and measuring the height of the incandescent zone in the bars after they are removed as previously discussed. Examples of the information gained from these rods include the following:

   - Ash depth from the width of lowest band on the rod. It will be light grey in color.

   - Heighth and location of the "fire zone" in the bed. This is evaluated by the location and width of the incandescent zone on the fire test rod. Sometimes the horizontal layering of zones in the gasifier bed is indicated by the appearance of two separate incandescent zones on the rod.

   - The physical condition of the bed is indicated to the operator by the "feel" of the bed during the insertion of the rods.

The possible deviations of these parameters from optimal operating conditions include:

1. Increase or decrease in the pressure drop across the bed.

2. Increase or decrease in the exit gas temperature.

3. Increase or decrease in the ash bed level.

4. Increase in resistance during bed probing.

1A. <u>Increase in Bed Pressure Drop.</u> An increase in the pressure drop across the bed can result from an increase in load, a change in the physical status of the ash in the ash bed, an increase in the ash bed level, agglomeration of coal across the bed at some point, a change in the size consist of the coal feed, and an influx of fines. Proper responses to these conditions would be as follows:

a) Check the saturation air blast temperature. If it is too high, reduce the temperature to a lower level. This will drop the moisture level of the air to the producer minimizing the chance of "mushiness" in the ash layer. Of course, known saturation temperature requirements, as dictated by ash fusion considerations, must be adhered to.

b) Check the height of the ash level. If too high, increase the grate speed to increase the ash removal rate, thereby lowering the ash level.

c) Check the condition of the coal bed by "poking" with the fire test rods. If part of the bed is found to be caking, increase the agitator speed and/or "poke" the bed vigorously where resistance is felt. Also an increase in the saturation temperature (more steam) should be considered.

d) Check the size consist of the coal feed. If large quantities of undersize are present, check the coal screening operations and make any required adjustments to return the coal zize consist to the proper quality.

1B. <u>Decrease in Bed Pressure Drop.</u> A decrease in the pressure drop across the bed is symptomatic of the opposite conditions discussed above. In addition, the formation of a "blowhole" is probable. A blowhole is a localized weakness in the bed which causes air flow to be channeled through a path of low resistance. This can be seen by visually inspecting the top of the bed. A blowhole will be recognized by a red glow seen at the top of the bed. A sharp rise in the CO_2 content of the gas generally accompanies the formation of a blowhole. An increase in the O_2 content of the gas further indicates the "breakthrough" of the air blast.

The gas off-take temperature will also rise, generally fairly sharply. To remedy this problem, the agitator speed should be increased to redistribute the coal around the top of the bed, filling in these weak spots and the load should be decreased until stable conditions are regained. "Load" is decreased by reducing the air flow.

2. Increase or Decrease in the Exit Gas Temperature.

Changes in the exit gas temperature are indicative of several operating problems. Should the temperature rise quickly, the cause could be the formation of a blowhole and/or a reduction in the saturation air blast temperature. If the increase is due to the formation of the blowhole, an accompanying decrease in the pressure drop across the bed will occur. Corrective action is the same as discussed above for blowhole problems. Should the saturation air blast temperature be too low, this could cause formation of clinkers in the system. The temperature should be readjusted quickly be decreasing the water flow rate through the jacket, which will cause more water to evaporate into the combustion air. A higher moisture content of this air stream will lower the temperature of the combustion and reaction zones. This will help prevent clinkering.

3. Increase or Decrease in the Ash Bed Level.

Changes in the ash bed level are never immediate in nature. An increase or decrease in the ash level normally results from increasing or decreasing the load or from a major change in the coal composition. The control provided is the grate speed. Simply increase the grate speed to increase the ash removal rate and vice versa. Of course, changing the "load", thereby consuming more coal, will change the rate of ash formation.

It is important to note that minor fluctuations around the desired operating parameters are to be expected, especially during load variation. Because of the high capacitance of the producer system, major upsets will not result from these minor fluctuations. Any minor deviation from the operating parameters for an extended period of time will, however, be difficult to correct in a short period of time. As an example, clinkers formed due to excessive bed temperature could accumulate, sometimes slowly, yet cause a shutdown if the problem becomes acute. Therefore, immediate corrective action should be taken when indicated.

4. Other Problems.

4.1 Emissions.

This is not a complete inventory of all emissions from the operation; only those that were the most troublesome.

a. Poke Holes. Copious quantities of gas can issue from the poke holes during inspection, rodding and during "fire tests" with the steel fire test rods.

Bureau personnel at Minneapolis developed an improved version of a South African poke hole design that largely eliminated this problem. Slight additional improvements to the poke hole cover should totally eliminate this problem. Other proprietary designs exist that are reputed to be equally effective. Those of you who know of this problem now know that a solution exists also.

b. Coal Feed Bins. The coal feed bins are under gasifier pressure during feeding of the coal. As the bins empty, gas fills the void created. Refilling the bins with coal displaces the gas to the atmosphere via the upper fill valve. To date, no simple economical method of coping with this problem has been found. Purging suggests itself, but no work has been done on this.

c. Dust. Dumping of coal into the open screening station bin at ground level resulted in much dust. The fuel dust was light and fluffy and deposited an annoying layer of dust on the lower levels of the structure. A commercial installation would have to resolve this problem.

4.2 Supplying Water Vapor for the Gasification Reaction.

As mentioned previously, the Wellman-Galusha gasifier supplies the water vapor (steam) necessary for gasification. This is accomplished by passing the air over the surface of the water in the jacket. When a certain sub-bituminous coal was fired, a solid wall of clinkers formed along the inner wall of the gasifier. This effectively shielded the jacket, and reduced heat transfer to the water in the jacket to such an extent that an auxiliary source of steam had to be used.

This is a very abnormal situation brought about by the gasification conditions used and the nature of the coal. This incident illustrates the importance of maintaining the "right" conditions for gasification.

SHUTDOWN

Whether a shutdown is planned or whether it is of the emergency variety, the producer is immediately isolated from the downstream system by opening the vent line valve and simultaneously stopping the air blower. Ideally, this action should be interlocked with a downstream controller. After the gasifier has been vented, the cyclone is filled with water thereby completely isolating the gasifier. To empty the gasifier, the ash removal grate speed is increased to maximum and the coal is discharged through the ash removal system. Precaution: It may be necessary to douse hot coals or remaining coals with water if the coal is highly reactive. The operation can be hastened by manual removal of the refuse through the ash pit doors. This is done when there is a need to get back on line in a hurry.

CONCLUSION

Operating experience has shown the Wellman-Galusha gasifier to be an extremely stable piece of equipment. If the key operating parameters are maintained at the required set points, the gasifier requires little other attention. Of most importance is determining those operating points which yield the maximum capacity of the gasifier with insured stability. This will, of course, depend on the coal to be gasified. To find these conditions with coals that have not previously been processed could require a testing program.

Specific information concerning the nature of the tests run this past fall at the U.S. Bureau of Mines will be forthcoming and may be published by the time of this paper.

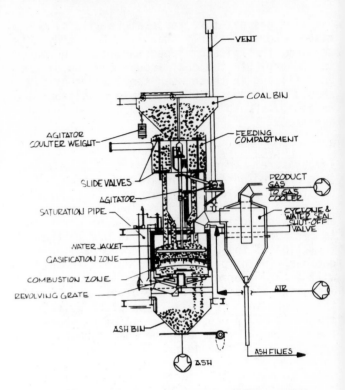

Figure 1. Wellman-Galusha agitated gasifier.

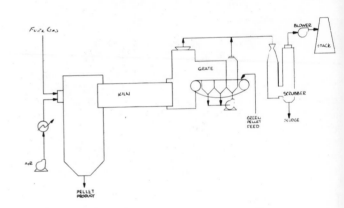

Figure 2. Pelletizing pilot plant.

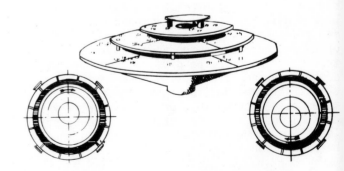

Figure 3. Wellman-Galusha rotary grate.

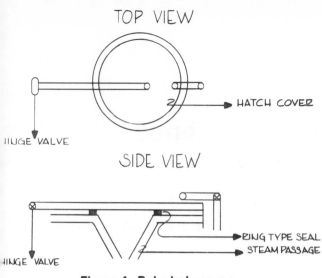

TOP VIEW

SIDE VIEW

HINGE VALVE

HATCH COVER

HINGE VALVE

RING TYPE SEAL

STEAM PASSAGE

Figure 4. Poke hole cover.

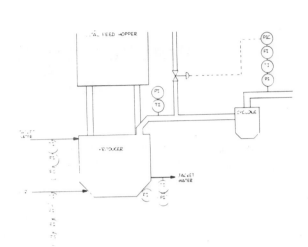

Figure 7. Placement of fire boxes.

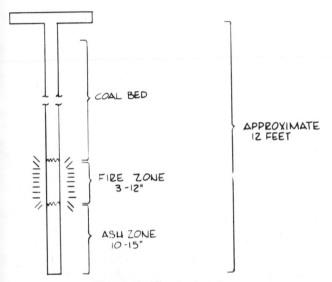

COAL BED

APPROXIMATE 12 FEET

FIRE ZONE 3-12"

ASH ZONE 10-15"

Figure 5. Fire test rod.

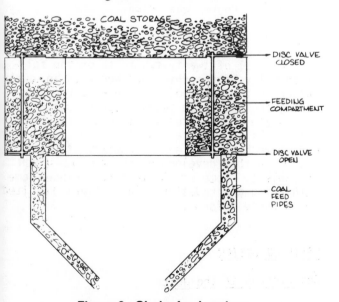

COAL STORAGE

DISC VALVE CLOSED

FEEDING COMPARTMENT

DISC VALVE OPEN

COAL FEED PIPES

Figure 6. Choke feed system.

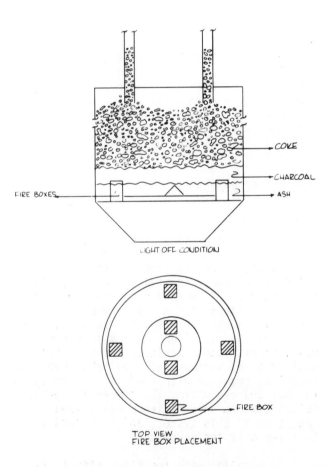

COKE

CHARCOAL

ASH

FIRE BOXES

LIGHT OFF CONDITION

FIRE BOX

TOP VIEW
FIRE BOX PLACEMENT

Figure 8. Control instrumentation.

Practical Operating Experience on a Riley Gasifier

This paper reviews the results of the operating history and design of a commercial-size atmospheric coal gasifier obtained over a five-year period.

W.P. Earley
W.P. Earley & Assoc.
West Brookfield, Mass.

R.A. Lisauskas and A.J. Rawdon
Riley Stoker Corp.
Worchester, Mass.

The need for gaseous fuels today parallels an earlier period in our industrial history (1,2, 3,4). The range of applications today is even wider and more diversified. The constraints of the environment and economics make filling this need even more of a challenge. This paper illustrates how that challenge has been met, with emphasis on actual operating results.

DEVELOPMENT PROGRAM

Background

In 1974, the Riley Stoker Corporation decided to make a considerable commitment to coal gasification, in light of the increasing evidence for the need of alternate fuel gases for industry. A two-phase program was initiated. The first was on a purely R & D level, and included the purchase of a two-foot (.6m) I.D. fixed bed gasifier. A complete process development system resulted, for the study and evaluation of various fuels suitable for gasification, together with a means for processing the gas and evaluating its combustion behavior. Valuable hands-on experience was gained in the operation of this system, and it served as a training unit for personnel.

In parallel with the pilot effort, Riley Stoker undertook a program of design, development and commercialization of a full-sized, low Btu coal gasification unit. During a peak period (1921), approximately 11,000 gas producers were in use by U.S. industry, (5) and that of this number a large portion were of a design developed by the Morgan Construction Company of Worcester, MA. In total, over 9,000 of Morgan units were manufactured and distributed throughout the world. They enjoyed a reputation for reliability, high capacity, and the ability to handle a wide variety of coals. An agreement was entered into whereby Riley Stoker obtained the exclusive manufacturing rights to the Morgan Gas Producer and a redesign program was begun.

Fuels Tested

A total of twenty-two tests on the full-sized Riley-Morgan unit together with an equal number on the two-foot unit were carried out during the past five years. Fuels ranging from sized anthracites to sized and run-of-mine coals having free swelling indices up to 8.5, have been studied. A summary of fuels tested in this development program is given in Table 1. Some of the test runs caused some rethinking and redesign. Each of them resulted in valuable information.

GASIFIER DESIGN

Description of the Unit

Figure 1 depicts the full-size 10'6"(3.2m) I.D. unit as it stands today. The major departures from other fixed bed processes are that it is a thin-bed, variable height process in which the entire fuel bed slowly rotates.

The thin fuel bed up to 55 inches (1.4m) in height allows variation in particle heating rate, the key to managing swelling coals. The rotating bed assures even distribution of fuel to the bed, an essential requirement when there is a size variation. This is accomplished by means of a rotating drum feeder, the length of which is nearly equal to the radius of the bed. Coal enters the reactor by falling through a slot across this radius, thus assuring a non-segregating feed to the top of the fuel bed with no moving internal mechanism. It must be noted that with thin fuel bed reactors, a variation of six inches (15 cm) in depth may represent a 25-50% variation in total active fuel bed height.

Fuel bed agitation, another important element in the gasification of caking coals, is accomplished by means of one or two water-cooled horizontal bars of approximately one radius in length each offset slightly from the center of the fuel bed. The bars are externally weighted and counterbalanced to achieve the desired depth of agitation. These bars act to prevent large agglomerates and open channels from forming in the fuel bed.

Figure 2 is a schematic of the gasifier and test system at Riley Stoker's Worcester, (MA) R & D facility. Air is supplied by a fan through an 18" (45 cm) I.D. pipe to the bottom of the gasifier. Steam is added through a sparger after the air flow meter and controlled on the basis of saturation temperature. The mixture is introduced into the bottom of the rotating pan through a central blast hood. While there is no grate in the system, the ash bed performs the function of a diffuser. The air-steam mixture moves countercurrent to the descending coal, first through the oxidizing zone, then reduction (gasification) zone, and the devolatilization zone.

Ash is removed from the bottom of the unit periodically by holding stationary a spiral arm on the floor of the unit. Relative motion between this arm and the rotating bed forces ash out to a water seal at the periphery. The ash is then lifted out of the water seal by sliding up a vertical chute or "ash plow." During normal operation the spiral arm and chute are allowed to rotate with the fuel bed.

The producer gas exits the building through a 36" (91 cm) I.D. refractory lined duct to a cyclone with dust drop, and then to a sampling and metering run. The gas is burned in either an enclosed flare stack on the roof of the facility or in a water-cooled test furnace.

Coal Feed System

The coal feed system is one of the areas where Riley Stoker has spent considerable effort in design improvement. Historically, most atmospheric gasifier manufacturers paid scant attention to fugitive emissions from the coal feed system. The main goal was getting fuel in. Consequently, each time coal was admitted to the reactor, some gas escaped into the coal delivery system. This resulted in most cases with main storage bunkers full of producer gas. Aside from the obvious hazard to personnel who might be called upon to work in this area, each of these bunkers was a bomb waiting to go off. Being too rich most of the time is probably what prevented explosions from occurring regularly. Most bunkers were, of course, vented to the atmosphere, a condition not to be tolerated today.

In the Riley Stoker lock-purge coal feed system, shown in Figure 2, a combination of gas tight lock valves, a lock hopper evacuation system and a nitrogen blanket combine to prevent any escape of gas from the system.

Table 2 gives analyses of the gaseous environment in the gasifier building and coal bunker measured during a Level 2 environmental assessment of our process. This table, together with Table 3, is presented to quantify fugitive emissions and not to determine worker health effects.

OPERATION

Start-Up

Over the past five years, the Riley Stoker start-up procedure has continually evolved. From bundles of oil-soaked waste piled on railroad ties, to spraying wood blocks with fuel oil, a system has been derived whereby pitfalls generated by such procedures may be avoided. It has been found that a careless start-up can have profound consequences during later stages of operation. This is perhaps of less importance in an installation which is being fired up to go on line and remain there for five years or more. Good start-up, however, is much more important to a test facility where all on-stream time is so valuable for data taking. Correction of poor start-up procedures is time-consuming and costly to a test program of limited duration.

Taking full advantage of the rotating bed feature, a fuel gas (LPG, MAPP, etc.) ignitor is used to fire a six inch (15 cm) layer of wood charcoal interspersed with a small quantity of split wood shingles. A rapid, smokeless light-off results and by the end of one hour, the bed is totally ignited and uniform, exit temperatures have risen rapidly, and coal admission may be initiated. Almost invariably, charcoal gasification has usually begun. The transition from full oxidation to partial oxidation takes place automatically as more fuel begins to react.

The key to the success of a fixed bed gasification process is the maintence of the classic profile which neatly delineates the various zones of activity. As mentioned elsewhere, achieving a level fuel bed is of utmost importance, especially in the case of thin bed gasifiers. The route to this state is by means of proper fuel feed distribution, rather than after-the-fact attempts to level the fuel bed by mechanical means. What is sometimes overlooked is the fact that mechanically attempting to redistribute non-uniformly sized particles on the fuel bed surface can lead to segregation. Relative motion causes the large particles to move readily while the smaller ones essentially remain in their original position as they fall through voids created by the coarser particles. Our tests have confirmed findings of earlier investigators (6) that it is clearly better to put the coal on the fuel bed surface correctly from the start rather than rely on mechanical redistribution.

One of the most commonly presented sets of data within the fixed bed is the familiar vertical temperature profile, showing a smooth transition from zone to zone. The radial temperature gradient particularly at the reactor wall can also significantly effect fuel bed permeability and overall gasification performance. Many operators in the past (7) have recognized that controlling wall effects was often the most challenging aspect of operating a fixed bed gasifier. The higher gas velocities, lower carbon inventory and higher temperatures at the wall lead to dominance of the exothermic gas phase reactions in this region, while a richer "core gas", low in CO_2 and high in combustibles exists away from the wall in the area where endothermic gas-solid reactions hold sway. Discrete sampling across an active bed has shown this to be true.

Such sampling, together with other purposes such as data taking (temperature measurements, fuel bed pressures) and corrective action (clinker removal or flattening of a severely distorted combustion zone), requires safe, clean access to the fuel bed. Riley Stoker has developed such a device which assures that migration of all gases is into the gasifier, rather than from it. An analysis of poke hole gas discharge measured during operation is given in Table 3, which illustrates this point.

FUEL PROPERTIES

Particle Size

It is surprising how little understood is the fact that all gasification processes are fuel size specific. There is no process existing which will gasify run-of-mine fuel today, and yet inquires are constantly received requiring just that of the fixed bed. But just as the introduction of fines into a packed bed of large sized particles in a fixed bed process adversely affects its specific gasification rate, so too would the introduction of six-inch lumps into a fluid bed.

From the large body of work done on fixed bed gasifiers, especially in England in the later years after such work had ceased in this country, it is evident that gasifier performance was first of all closely linked to fuel sizing. Carefully graded fuels were becoming more expensive and difficult to come by. Substitution of fuels containing higher proportions of fines 1/8" to 3/4" (0.3 to 1.9 cm) began to result in decreasing unit outputs and units requiring more manual care, all other elements remaining constant.

Economics aside for the moment, and given a choice, the preferred size range for most fixed bed producers is 2" x 1" (5x2.5 cm) with some extensions being made (usually with a capacity penalty) to 3/4" (1.9 cm) or 1/2" (1.3 cm).

Experiments with nut-sized 2¼" x 1¼" (5.7 x 3.2 cm) and pea sized 5/8" x 3/8" (1.6 x .95 cm) anthracite have shown a decrease of 38% in specific gasification rate for the smaller fuel. Others have shown decreases of as much as 75% in specific gasification rates with the addition of a little as 25% fines below 1/8" (.3 cm) to a graded 1½"x 3/4" (3.8 x 1.9 cm) fuel, (8). Our test experience on both the smaller process development unit and full-scale Riley unit confirms hese findings.

The sensitivity to size, of course, is elated to fuel bed permeability. With increasingly smaller particle sizes, bed pressure drop increases, weak points become suscepible to channeling, temperatures within the ed become more non-uniform, and a whole cycle f undersirable occurrences takes place.

A much wider range of sizes could be olerated if a) size could be uniformly disributed within the range, and b) this distribation could be maintained within the bed. Unfortunately, such a result is further complicated by two other factors in handling of any solid fuel:

1. Friability of fuel
2. Segregation and breakage by fuel handling equipment.

Either one or both of these factors can entirely negate careful and costly fuel sizing and should be fully understood by the process engineer.

welling and Agglomeration

The time-temperature history of large particles in a fixed bed gasifier is significantly different than that of finer particles in either a fluidized bed or an entrained uspension gasification process. In a fixed ed reactor pyrolysis is initiated at the top of the fuel bed under a reducing atmosphere in a decreasing temperature gradient. Variables such as temperature, heating rate, particle size and both coal and gas properties all strongly affect the devolatilization process in the upper fuel bed. Understanding the behavior of particles during devolatilization is essential in avoiding the wrong application of leveller or agitator mechanisms. This is true for both caking and non-caking coals.

The devolatilization of a small number of particles can be studied in a simple retort under simulated gasifier exit conditions. The results of such a test on an eastern bituminous coal with a free swelling index of 4 1/2 and a non-swelling northern plains lignite is illustrated in Figure 3.

In each test three pieces of sized fuel 1" x 3/4" (2.5 x 1.9 cm) were inserted into an oven preheated to a desired control temperature. The coal particles were arranged touching each other and a blended producer gas mixture was fed into the oven chamber. The object of the experiment was to simulate the heating rate experienced by large coal particles falling onto a gasifier fuel bed. After devolatilization was complete the char particles were removed, weighed and then tested for strength in a drop shatter test.

It may be seen in Figure 3 that the swelling for each group of bituminous coal particles was not the same. Less swelling and less surface flow appears to have occurred as the temperature was increased. At high heating rates a steep temperature gradient is produced throughout the large coal particle. Under these conditions the outer layer of the particle exists in a plastic and liquid state for only a very short period. An outer semi-coke shell is formed before a deep plastic layer develops. This shell is strong enough to restrict further expansion of the particle. At lower particle heating rates temperature gradients are much less steep. In the experiment described by Figure 3 a large agglomerated mass was formed at a temperature of 750°F (400°C). This mass had the fragility of a Christmas Tree ornament.

Unlike bituminous coal the lignite particles did not noticeably change in volume when heated nor did they fuse with adjacent particles. The particles appeared to exhibit a distinct laminar structure with splintering occurring along the bedding planes.

The effect of temperature and heating rate on the strength of lignite char was found to be directly opposite to that for bituminous char. The amount of lignite char breakage in a drop shatter test was found to increase with higher retort temperatures while the amount of bituminous char breakage decreased.

In summary, for each coal, there is an optimum temperature and particle heating rate condition in the upper fuel bed. Proper management of this temperature together with mechanical agitation will effectively avoid caking difficulties associated with swelling coals.

Ash Fusion Temperature

The behavior of ash in the fuel bed and its separation and removal from the gasification reactor is an integral part of the gasification process. The fusion and clinkering of ash are important concerns in fixed bed gasifier operation. Ash fusion temperature is only one indicator of any fuel's clinkering temperature. Other factors known (9) to influence this tendency are the quantity of ash, fuel size and bed permeability, gasification rate, and fuel reactivity. If locked into the use of a fuel with a low ash fusion temperature, a full scale trial is the only true answer to its clinkering behavior. A hedge could be made by building spare capacity into the plant and later running with higher air-steam ratios to the optimum point.

The optimum point for a "dry bottom" gasifier is, of course, controlled clinkering, the production of small, granular pieces of fused ash. Most operators will tend to run on the conservative side of this condition, because it is a sensitive balance point. A danger in erring on this side, unfortunately, is the possibility of producing too fine an ash, which may adversely affect the flow pattern of gases within the full bed. Certain Western coals with significant concentrations of natural fluxes (Ca, Mg, Na) may produce extremely fine ash in the form of cemented particles rather than fused masses. Coal rank and ash chemistry, therefore, are important factors.

OPERATING RESULTS

Operating test conditions and results for a full scale Riley gasifier are summarized in Table 4 for an eastern bituminous coal and a northern Great Plains lignite. The much lower ash fusion properties of the lignite are counteracted by its higher reactivity. Only a slightly higher input steam-air ratio is required on lignite to control fuel bed temperatures. The high moisture content of lignite is reflected in the much lower gasifier outlet temperature for this fuel. Thin bed operation is an advantage in dealing with both the caking bituminous coal as described earlier and a high moisture coal as well. It allows a gasifier outlet temperature high enough to prevent early condensation of aqueous liquor.

The tar yield is considerably higher for the caking bituminous coal as compared with lignite even though both fuels are high in volatile content. This result is consistent with our earlier work (10) in which we attempted to evaluate gasifier tar yields for various fuels. Full scale Riley gasifier tar yield data is compared with yields from laboratory pyrolysis yield data in Figure 4.

The impact of tar yield is also reflected in a comparison of observed gasification efficiencies for each fuel type given in Table 5. There is significant difference in efficiency

when the heating value of tar and oil vapors in the raw gas are not considered. The 9% difference is considerable. This difference disappears and actually reverses when tar and oil are included in the efficiency determination.

Typical raw gas compositions on a saturated standard gas base are also compared in Table 6 for each fuel. Differences in volatiles evolution and the reactivity of each fuel are reflected by each gas composition. The gas produced by eastern bituminous coal is higher in hydrocarbon gases while the gas produced from lignite is higher in carbon-monoxide and hydrogen and lower in carbon dioxide content. The heating value and carbon monoxide-hydrogen ratio of the gases are also different. It is clear from these results that an understanding of devolatilization for each fuel type is necessary in order to describe the heating value and composition of the product gas as well as tar yield.

The gas composition results also indicate that thin bed operation does not result in a poor gas quality.

SUMMARY

Fixed bed gasifiers have the following advantages:
1. Long particle residence time that permits almost complete carbon conversion
2. Good particle-gas heat exchange, due to countercurrent flow
3. Minimal dust carryover
4. An inherently simple control philosphy
5. Large carbon inventory.

The following areas are of a primary concern:
1. Caking properties of fuel
2. Sized fuel required for maximum output
3. Agglomeration of ash
4. Permeability and uniform flow resistance of the bed
5. Tars and other by-products.

The following are some operating and design principles which we have found must not be violated, unless the trade-offs to capacity, smoothness of operation and decrease in efficiency are recognized and are acceptable:
1. Careful sizing is a must for maximum throughput, i.e. (2x1) screened and

washe...

2. For... an optimum exit
   tem... can be
   go... ...ral,
   th...
   s...

3. C...

4. ...

5. ...

6. ...

REFER...

1. ...

...on

2. ...

...s:

...N,

...'

..., Pa.,

...Graw Hill,

...U Industrial

Fuel Gas (Producer Gas) and Industrial Hydrogen," Battell Memorial Institute Final Report to OCR, January, 1977.

6. Burke, S. A. and Sparham, G.A., _Coal Utilization Research Association_, London 16 (8), August, 1952.

. BCURA, "Proceedings of the Conference on Practical Aspects of the Generation and Utilization of Producer Gas," Harrogate, Yorks, April, 1957.

8. Von Fredersdorff, C.G., and Elliott, M.A., _Chemistry of Coal Utilization_, Supplementary Volume, Ed. H. H. Lowry, p. 951ff., John Wiley & Sons, N.Y., 1973.

9. Hebden, D., "High Pressure Gasification Under Slagging Conditions," A.G.A./ERDA/IGV Seventh Synthetic Pipeline Gas Symposium, Chicago, 1975.

10. Lisauskas, Johnson and Earley, "Control of Condensable Tar Vapors From a Fixed-Bed Coal Gasification Process," Fourth Energy Resource Conference, Lexington, KY., January, 1976.

11. Rhodes, E.O., _Chemistry of Coal Utilization_, Volume 1, p 1281 ff., John Wiley & Sons, N.Y., 1945.

EARLEY, W.P.

LISAUSKAS, R.A.

RAWDON, A.J.

Table 1. Coals tested in Riley Stoker gasification test facilities.

COAL	RANK	NOMINAL SIZE	FREE SWELLING INDEX	ASH FUSION TEMPERATURE, °F (FLUID-REDUCING)
2 FT. DIAMETER PROCESS DEVELOPMENT UNIT				
Anthracite, Pa.	AN	PEA	0	2700
Pocahontas Seam, Va.	LVB	3/4" × 1/2"	3	2700
Sewell Seam, W. Va.	HVAB	2" × 3/4"	8	2700
Egypt Valley, Ohio	HVCB	2" × 1/4"	4	2290
Illinois No. 6	HVCB	BRIQ	2.5	2160
Northern Plains	Lignite	2" × 3/4"	0	2100
		2" × 1/2"		
		2" × 1/4"		
		ROM		
		BRIQ		
RILEY DEMONSTRATION PLANT				
Anthracite, Pa.	AN	NUT & PEA	0	2700
Upper Banner Seam, Va.	HVAB	1 1/4" × 1/4"	6	2630
Coronet No. 2, Va.	MVB	2 1/2" × 1"	8.5	2560
Hazzard No. 4, Ky.	HVAB	2 1/2" × 1"	4.5	2700
Elkorn No. 3, Ky.	HVAB	2" × 1 1/2"	4.5	2660
Northern Plains	Lignite	2" × 1/2"	0	2050

Table 2. Gasifier plant environment organic vapors analysis.*

LOCATION	CONCENTRATION (ppm as CH₄)
GASIFIER BUILDING WALKWAYS	<1.0 ppm
TOP OF GASIFIER DURING POKING OPERATION	≤1 ppm
COAL BUNKER	5-6 ppm

* Data collected by Radian Corp. under EPA contract to provide data for environmental assessment of low Btu gasifiers.

Table 3. Riley gasifier measured poke hole gas discharge during operation.*

GAS	VOL. %
N_2	95.4
H_2	1.1
O_2	0.2
CO	Below detection limit
CH_4	Below detection limit
CO_2	Below detection limit

Data collected by Radian Corp. under EPA contract to provide data for environmental assessment of low Btu gasifiers.

Table 4. Summary of operating conditions and results related to coal data.

	EASTERN BITUMINOUS	NORTHERN PLAINS LIGNITE
Coal Data		
Size (Nominal)	2" × 1 1/2"	2" × 1/2"
Moisture (As Received) Wt. %	4.3	32.8
Ash (dry) Wt. %	3.9	9.8
Volatile Matter (dry) Wt. %	41.1	42.0
Fixed Carbon (dry) Wt. %	55.0	48.2
Higher Heating Value (dry) BTU/lb	14570	10760
Ash Softening Temperature. F		
(Reducing H = 1/2W)	2600	2020
(Oxidizing H = 1/2W)	+2700	2190
Free Swelling Index	4.5	0
Operating Conditions		
Air. lb/lb daf coal	3.11	2.44
Steam/Air Ratio wt/wt	0.14	0.18
Fuel Bed Height. inches	46	48
Operating Results		
Outlet temperature. F	1292	518
Gas Yield SCF/lb daf coal	69.3	56.1
Gas Heating Value BTU/SCF	156	166
Tar Yield lb/lb feed coal	0.087	0.02
Moisture lb/lb feed coal	0.25	0.44

Table 5. A comparison of gasification efficiencies with coal type.

GASIFICATION EFFICIENCIES		EASTERN BITUMINOUS	NORTHERN PLAINS LIGNITE
(a)	$\dfrac{\text{Heating Value of Raw Gas} \times 100}{\text{Heating Value of Gasified Coal}}$	71.4	78.0
(b)	$\dfrac{\text{Raw Gas} \times 100}{\text{Gasified Coal + Steam}}$	69.0	74.9
(c)	$\dfrac{\text{Raw Gas + Tar + Oil} \times 100}{\text{Gasified Coal + Steam}}$	78.3	77.9

Table 6. Comparison of raw gas composition with coal type.

GAS COMPOSITION (VOL. %) SATURATED AT 60 F, 30" Hg	EASTERN BITUMINOUS	NORTHERN PLAINS LIGNITE
CO	21.6	28.1
CO_2	7.5	6.1
H_2	13.9	17.3
CH_4	2.2	1.5
C_nH_m	0.9	0.2
COS & H_2S	0.1	0.1
Inerts	0.6	0.5
N_2	51.5	44.5
H_2O	1.7	1.7
	100.0	100.0
Higher Heating Value (Btu/ft³)	156	166
CO/H_2 Ratio (Vol/Vol)	1.55	1.62

Figure 1. Riley gasifier cross-section.

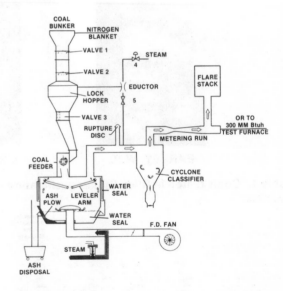

Figure 2. Schematic of Riley Stoker coal gasification demonstration plant.

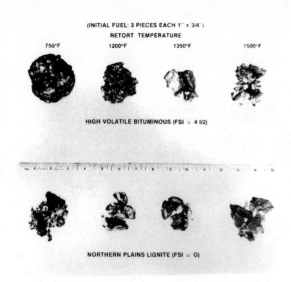

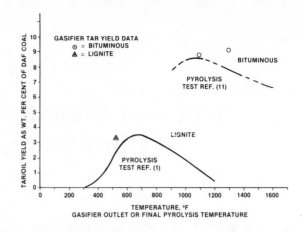

Figure 4. Comparison of gasifier tar yield with "coal" pyrolysis test data.

Figure 3. Char residue devolatilization test.

Hot Gas Cleaning for Advanced Coal Conversion Technologies

An effort was made to present the problems associated with many advanced coal conversion technologies.

D.F. Ciliberti, A.Y. Ranadive, D.L. Keairns,
P.R. Mulik, and M.A. Alvin
Westinghouse R & D Center
Pittsburgh, Pa.

INTRODUCTION

One of the most pressing problems facing the United States today is the shortage of economical domestic petroleum and gas. This has caused an unsettling degree of dependence on foreign sources, and the difficulties have been compounded by recent problems with the acceptance of nuclear power production. These factors have given renewed impetus to efforts in the development of new and better methods for utilization of the nation's coal reserves.

To ensure implementation of any of the new coal conversion technologies for electric power production, two basic criteria must be met:

- The method must be environmentally safe and acceptable

- The method must successfully compete with the economics of conventional power plants.

Perhaps two of the most promising new near-term techniques for coal-generated electricity is the concept of pressurized fluidized bed combustion (PFBC) and fluidized bed coal gasification.

Typically, these processes employ a combination of the steam turbine cycle and the gas turbine cycle (in which the hot pressurized products of combustion from the bed are expanded through a gas turbine).

The economic advantage of these processes over conventional coal-fired plants with stack gas scrubbers is on the order of a 5-8% increase in overall plant efficiency.

For pressurized fluidized bed combustion, cleaning the gases at temperatures and pressures on the order of 800-900°C and 10 to 15 atm. prior to expansion in the turbine is essential. For gasification (especially the low-Btu schemes) hot gas cleaning represents a significant potential for higher plant efficiency over more traditional cold gas scrubbing techniques, since a considerable portion of the energy of the product gas is represented by the sensible heat of the gas.

As with many coal conversion technologies, the success of these techniques hinge in large part on their ability to remove the contaminants of coal combustion to the required degree. Current legislation limits the amount of SO_2 which may be released to the environment by such processes. It is, therefore, not surprising that a great deal of research and development has been expended on methods of sulfur control. Indeed, the great capacity of these systems to deal effectively with sulfur constitutes one of the attractions of these processes. Similar legislation for the total mass of particulate emissions has been promulgated; however, the degree of success in controlling particulates in high temperature and pressure environments has not been commensurate with either proposed environmental standards or standards based on

0149-3701/81/4174 $02.00 © 1981 AIChE

conomic turbine life. Particulates, then, ot only cause a still unresolved impact on he environmental acceptability of PFBC and asification; they may also have an impact on he operating cost and availability of combined ycle plants.

Alkali metal emissions from PFBC and gas-fication are not likely to entail significant nvironmental effects; however, they can have large impact on the reliability and mainten-nce costs of combined cycle plants. Although ot corrosion and fouling of critical turbine omponents by alkali metal compounds is a ajor technical concern, hot pressurized emoval of alkali metal compounds has received ittle attention by comparison with removal of articulates and SO_2. As a result, there are o proven methods for dealing with the problem.

he Particulate Control Problem

In discussing particulate control as it elates to PFBC and gasification, it is help-ul to divide the overall problem into three reas. The first deals with the problem of etermining what particulates are presented to he cleanup system. The second deals with he dual acceptability problem of determining urbine tolerance to particulates, as well as roper environmental specifications. These onsiderations set the goals of the final area f endeavor, that of determining the means by hich the levels of particulates presented to he clean-up system are reduced to what has een determined to be acceptable levels.

articulate Carry-Over

Particulates entrained in the hot pres-urized gas from the bed can arise from three ources: unreacted char, coal ash, and (for FBC) small particles of the limestone or olomite sulfur sorbent. The importance of otal systems considerations on particulate arry-over should not be overlooked. Factors uch as sulfur sorbent choice, size, method of creening, and degree of sulfur removal can ave a significant effect on the particulate urden. Similarly, the choice of operating onditions can produce an enormous effect on arry-over. Choices of superficial velocity, ed height, and whether or not to recycle ines are critical determinants of what is resented to the cleanup system.

Over the past few years, Westinghouse has volved computer models of the PFBC process hat allow a qualitative assessment of the ffects of such parameters. With these tools,

the effects of any parameter on the particu-late carry-over, as well as the loadings and size distributions of particulates throughout the PFBC system, can be determined quickly and easily. These tools have been used to examine the impact of stricter SO_2 emission limits on particulate loading (1); it was determined that an increased Ca/S ratio would cause much higher particulate loadings due to increased attrition in the bed. Figure 1 presents a representative projection of particle profiles through a typical PFBC system in the fines recycle mode. Figure 2 presents a projected size distribution for the Exxon miniplant emissions, along with the range of typical data for comparison (2). On the basis of these and several other such exercises, some apparently general descriptions of the parti-culate presented to the final cleaning stage can be made. First, the particulate pene-trating two stages of conventional cyclones is expected to be approximately log normally distributed between 0.1 and 10 to 20 μm, with an average diameter of about 1 to 4 μm (as indicated in Figure 1). Second, the loading will most likely be on the order of 0.5 to 1 gr/scf, although this quantity is subject to variations deriving from its sensitivity to different design and operating philoso-phies. Current efforts with this modeling center on improved methods for handling attrition and elutriation effects so that better correlation with operating systems can be made.

Particulate Effects in Gas Turbines

Over the past 10 years, Westinghouse has been developing its own models as tools for determining particulate control standards and acceptable turbine operation with coal-derived fuels. As a major manufacturer of utility gas turbines, Westinghouse has developed sub-stantial experimental and analytical resources to evaluate gas cleanup requirements and turbine performance for operation with PFBC power plants. Among these resources are a turbine erosion damage model and a turbine deposition model.

These models are currently being applied on an EPRI program, where the objective of this effort is to provide improved estimates of PFBC power plant gas cleanup requirements for acceptable gas turbine service. The erosion and deposition models are being used to determine particulate removal requirements, by establishing turbine degradation as a function of gas cleanup. The empirical input data (particle erosivity and extent of

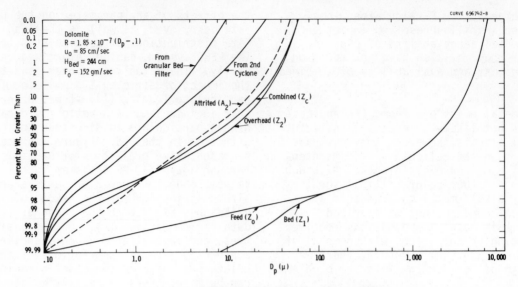

Figure 1. Particle size distribution profiles.

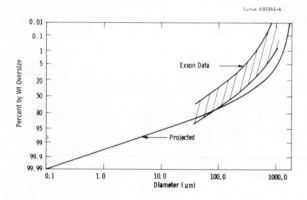

Figure 2. Total solids from combustor (PFBC).

adhesion) to the models is currently being improved so these models can reach their full potential.

The erosion model has successfully been coupled with the particle carry-over program so that economic trade-offs between extensive particulate control measures and turbine blade life have been assessed (3).

Deposition model predictions have been compared to experimental results. Experimental data for the particle size range appropriate to turbines is available for flow in pipes. Agreement of the deposition model predictions with this data was found to be excellent, as was the agreement between model predictions and measured deposition in turbine cascade experiments (4).

Tolerances and gas cleanup requirements are currently being evaluated for two large utility turbines operating with PFBC power plants using the erosion and deposition models described above and experimental results from recent EPRI and DOE programs. Although these evaluations have not been completed, preliminary assessments indicate the particle loading and distribution that would result in a 10,000 hour blade lifetime due to erosion may fall in the range of 0.004 grains/scf to 0.02 grains/scf for a mass median particle diameter of about one micron. Preliminary assessments indicate that the maintenance interval to remove deposits and thereby limit drops in turbine power to less than 10% may fall in the range of 100 to 500 hours for the above particle loading and distribution. The deposits do not appear to be tenacious so that removal should readily be accomplished by standard methods, such as water washing, during normal maintenance, and as such the deposits do not appear to represent a serious operating problem.

Particulate Control Devices

Unlike the carry-over and turbine tolerance work the majority of our current work in particle collection device characterization has been experimental in nature. Current testing programs include high temperature and pressure testing of a rotary flow and conventional cyclone and two types of ceramic fiber bag filters (both a woven cloth and a felt or fibrous type). Additionally, a DOE-sponsored program is just starting in which bench scale testing of a rigid ceramic membrane filter will be conducted. All this work (with the exception of the ceramic membrane filter) is being carried out at our high temperature and pressure particulate control test facility.

This facility has the capability of testing equipment on a large scale at operating pressure and temperature. The system characteristics are listed in Table 1.

Table 1

High-Temperature and -Pressure System

Temperature	- to 875°C (normal operation 760°C)
Pressure	- to 1380 kPa (normal operation 1034 kPa)
Gas Rate	- design to 30 m³/min at 1034 kPa and 875°C
Dust Loading	- virtually any desired
Dust Size	- virtually any desired
Combustor	- standard 15-cm gas turbine
Air	- preheated to 540°C
Fuel	- No. 2 diesel
Piping	- 25 cm with 5-cm insulation and 15-cm Inconel liners
Vessel	- 1.5-m diameter with 5-cm insulation
	- 1.8 m straight side
	- 4-25-cm nozzles
	- 3-7.6 cm nozzles

A schematic of the facility is presented in Figure 3. The piping and valving arrangement of the system is flexible enough to allow virtually any particle removal device to be tested. Under the sponsorship of Westinghouse, DOE, and EPA, the passage has been erected and shaken down.

Rotary Flow Cyclone Tests

Efforts to characterize an advanced rotary flow cyclone's performance has been a continuing project which has been sponsored by the Department of Energy. Previous results have been reported as the data have become available (5,6,7). The most recent tests were performed on an 11.4 m³/min. unit (11.4 m³/min. dusty air through the primary and 7.5 m³/min. clean air through the secondary). A series of runs were conducted at total pressures of 11 atm. and temperatures of 260°C, 540°C, and 815°C. These tests were carried out using a standard test dust (AC fine) which had an average particle size of about 3 μm with 95 wt % of the material being smaller than 20 μm as measured by cascade impactor. Figure 4 presents the results of the 815°C runs in the form of a grade efficiency curve. The grade efficiency and over-all efficiency (about 84%) were determined by simultaneous inlet and outlet sampling with cascade impactors. The pertinent pressure drop data for the system were as follows:

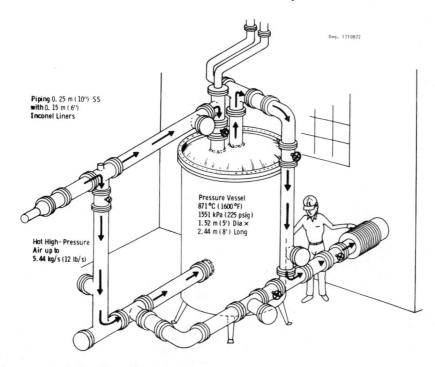

Dwg. 1710B72

Piping 0.25 m (10") SS with 0.15 m (6") Inconel Liners

Pressure Vessel
871°C (1600°F)
1551 kPa (225 psig)
1.52 m (5') Dia ×
2.44 m (8') Long

Hot High-Pressure Air up to
5.44 kg/s (12 lb/s)

Figure 3. High-temperature and - pressure particulate removal test facility (split inlet).

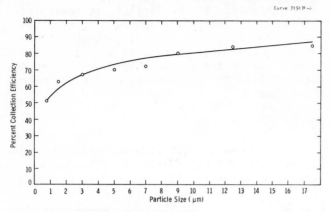

Figure 4. Grade efficiency curve for rotary flow cyclone (815° C, 11 atm, AC fine).

Dust hopper to outlet - 27.7 cm wg

Primary inlet to outlet - 33.8 cm wg

Secondary inlet to outlet - 38.9 cm wg

Within the accuracy of the measurements no significant effect of temperature on collection efficiency was observed. Pressure drops varied predictably with the gas density. The cut points measured in these experiments varied in the range of 0.5 to 3 μm. It should be noted that the data at these smaller sizes were quite variable. One peculiarity consistently observed was the significant penetration of some larger particulates in the 15-20 μm size range. The penetration of these larger particles could have serious implications for turbine performance and life in systems that contemplate the use of such cyclones for the final stage of cleaning.

High Temperature and Pressure Bag Filter Test Program

Because of the stringent cleaning requirements of gas turbines and the ever tightening environmental controls on fine particulate emissions, it seems likely that some form of positive filtration will be required. Accordingly, under the sponsorship of EPRI, we are currently carrying out a test program to evaluate two novel concepts in high temperature filtration. The devices being tested are both bag filters made from ceramic fibers. One filter is a woven cloth type and the other is a fibrous mat supported between screens. The test program as currently conceived is more of a feasibility assessment than a detailed, all encompassing experimental program. Test conditions for these experiments nominally are as follows:

Temperature - 250-825°C

Pressure - 11 atm

Flow Rate - 7-30 m^3/min.

At this writing only preliminary data on the woven bag filter are available. Figure 5 presents the results of a 425°C, 11 atm. test using a mixture of 75% Exxon miniplant flyash collected in the second cyclone with 25% of a finely ground (≤ 10 μm) dolomite dust. Pertinent observations to date include:

- high overall efficiency (99 + %)

- modest pressure drop (about 2 cm wg for the newly cleaned bag at face velocities of about 2 m/min.)

- successfully cleaned by momentary back pulse of high pressure air.

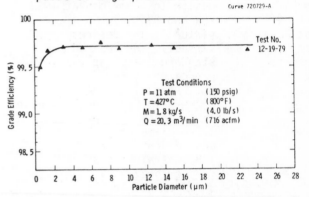

Figure 5. Summary of woven ceramic bag filter test results at air to cloth ratio of 1.7 m/min (5.6 ft/min).

Questions of bag life and durability cannot be answered in full by this test program, but the initial results are certainly encouraging.

Alkali Control Problems

The potential for turbine corrosion in PFBC and gasification application results from the formation of alkali phases that can then be carried through the system and deposited in the turbine. These released alkali phases (if not removed in the hot gas cleaning system) may lead to corrosive attack from either gaseous anionic transport of the metals which react with sulfur oxides to form sulfate deposits on the turbine blade, or from transport on particulate surfaces forming "sticky" corrosive agglomerates. Similarly, any vanadium present during combustion will form alkali vanadates and vanadium pentoxide. Sodium sulfate, sodium vanadate, and vanadium pentoxide are semi-molten and corrosive at metal temperatures as low as 620°C (1150°F). They become severely corrosive at the operating temperature of the pressurized fluidized bed.

The release of volatile alkali and trace metal compounds from coal, dolomite, or limestone used in advanced coal conversion technologies depends largely on (1) the distribution of alkali or trace metal containing phases in the feed materials, (2) the temperature and the local chemical environments that these phases are exposed to, (3) the time of the reaction, and (4) the competitive rates of different thermochemically favored reactions. The high volatility of sodium chloride at projected process temperatures makes it likely that the coal or sulfur sorbent that contains entrapped saline waters will release some sodium (as the chloride) into the gas phase. This alkali has been projected to be released at approximately one percent of the initial feed content, due to the simultaneous interaction of the clay fraction which reacts to form non-volatile alkali aluminosilicate or silicate phases.

Efforts to define acceptable levels of alkali have centered on the thermodynamic projection of conditions that would allow condensed alkali phases to exist on metallic components of the turbine. The thermodynamic conditions for deposition of sodium sulfate from fluidized-bed combustion effluent gases are shown in the 3-D representation of Figure 6. Gas phase compositions which result in the formation of liquid sulfate can be avoided by lowering the metal temperature and system pressure, lowering the sodium concentration, or increasing the excess chlorine (as HCl) concentration. In the 3-D projection, it is obvious that release to the gaseous phase of even a small fraction of the sodium or potassium in coal is sufficient to exceed the turbine tolerance of alkalis. Efforts such as these have led us to believe that it is likely that successful operation of a coal based combined cycle plant will require some form of alkali metal control, be it in conjunction with particulate control, or in-bed control or by turbine blade protection.

An apparently attractive option would be to combine final filtering for particulate control with a stage of alkali metal removal. This might be effected by the use of a granular bed filter composed of alkali sorbent, or by injection of sorbent powder into the hot gas stream ahead of the final filter.

Current Experimental Program

Under the sponsorship of DOE, we have initiated an experimental program to investigate some of these options for gasification systems. The goals of the program are to screen potential sorbent materials; determine

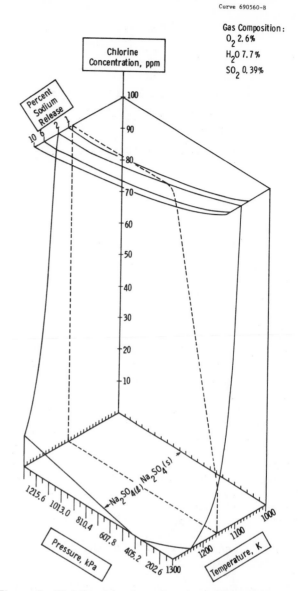

Figure 6. Fluidized-bed combustion effluent gas composition projected to cause sodium sulfate deposition.

required kinetic and other design data; and finally to conceive of methods for implementation of the process into actual equipment. It should be noted that a similar program for fluidized bed combustion systems is currently being actively pursued at Argonne National Laboratory (8).

To date our experimental efforts have centered on the identification of potential sorbent materials and subsequently exposing them to alkali laden gases in the high temperature and pressure flow reactor shown in Figure 7 to determine relative potential for alkali "gettering". With this system it is possible to insert and withdraw "plugs" of

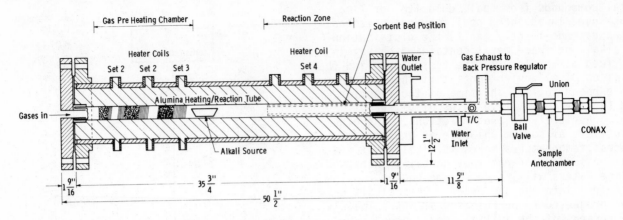

Figure 7. Bench scale reactor for preliminary high pressure-temperature studies of alkali sorption.

pelletized getter materials while the system is at temperature and pressure.

Materials exposed in this apparatus are then submitted for chemical analysis which consists of the following: atomic absorption for wt % of soluble and insoluble alkali, x-ray diffraction for crystalline compound identification; microprobe indicating surface and radial alkali concentration profiles and BET for surface area estimates. Figure 8 presents the results of a typical run in this apparatus. During these runs, cylindrical pellets of kaolin (.3 cm diameter by .3 cm long) were exposed to alkali-laden gases for periods up to 4 hours at temperatures of 850°C and 900°C while at a total pressure of 11 atmospheres. The gas phase consisted of 115 ppm NaCl (850°C) and 205 ppm NaCl (900°C) in Argon; the gas was regulated to a flow rate of 14 liters/min. (standard conditions) through the 2 cm diameter reaction tube.

Exposure of the 850°C and 900°C kaolin pellets to the sodium chloride vapor for four hours increased the total sodium content within the aluminosilicate 12.5 and 13.3 fold, respectively, over background. As shown in Figure 8, the lower reaction temperature maintains the alkali within the soluble fraction, whereas at 900°C the insoluble sodium complex is the preferred stable phase. The corresponding trend is shown in the background potassium content of kaolin. Although potassium is not included in the reaction gas mixture, the preliminary data illustrate the effect that time and temperature have on the potassium background - that is to volatilize some fraction of the original background potassium and release it to the gas stream. The inability to hold alkali at high temperatures and low gas phase concentration can have serious implications

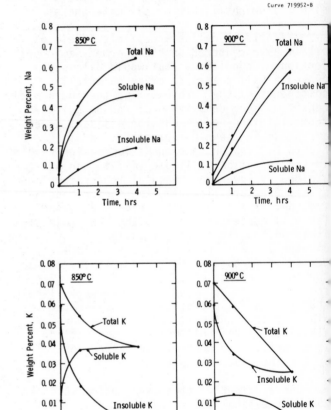

Figure 8. The effect of temperature and time on alkali reaction in kaolin in argon at 10 atm. pressure.

for the attainment of the very low alkali concentrations required.

Another aspect of this program has been the design and fabrication of a high temperature and pressure apparatus that is capable o providing the detailed single particle kineti data on alkali gettering that will be required for design purposes.

One of the major problems encountered in carrying out these experiments is the measurement and control of the gas phase alkali concentration. Because the vapor pressure curves are so steep at these temperatures an error of 50°C can cause a two-fold difference in alkali concentration. Adequate on-line instrumentation for monitoring the gas phase alkali concentration at temperature and pressure could not be identified, so a direct gravimetric method was opted for. A schematic of the test apparatus is shown in Figure 9. The system is basically a double balance thermogravimetric system employing two Cahn 1000 balances - one to continuously monitor the alkali source weight loss and the other to monitor the weight change of the candidate sorbent pellet. With this system, continuous curves of alkali sorption will be generated, which when coupled with precise knowledge of the gas phase alkali concentration will permit detailed kinetics of the gettering process to be determined.

Summary

In this paper an effort was made to present a broad overview of the problems associated with high temperature and pressure particulate and alkali control for advanced coal conversion technologies that utilize the combined cycle concept. Additionally, a brief discussion of the goals and preliminary results of several Westinghouse programs in this area has been presented. It is our belief that although proven definitive solutions to these hot gas cleaning problems do not exist at this time, current programs such as ours and those being conducted by others active in the field will result in technically and economically viable solutions in the near term.

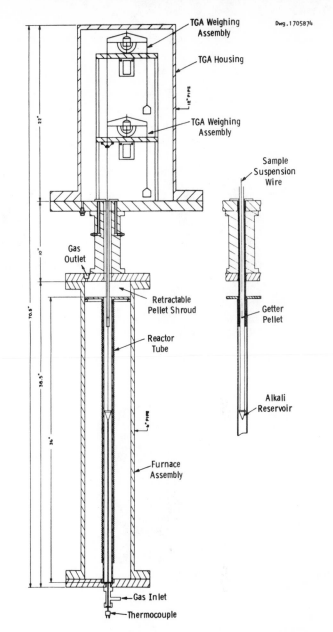

Figure 9. Proposed two balance system for kinetic studies of Gettering Process.

LITERATURE CITED

1. Newby, R. A., N. H. Ulerich, D. F. Ciliberti, and D. L. Keairns, "Effect of SO_2 Emission Requirements on Fluidized-Bed Combustion Systems; Preliminary Technical/Economic Assessment," EPA-600/7-78-163, NTIS No. PB 286971 (1978).

2. Brooks, R. D. and J. R. Peterson, Proc. of 5th Intl. Conf. on FBC, Washington (1977).

3. Keairns, D. L., D. F. Ciliberti, A. Y. Ranadive, J. R. Hamm, R. W. Wolfe, and E. F. Sverdrup, 24th Intl. Conf. on Gas Turbines, San Diego (1979).

4. Parker, G. J. and P. Lee, Proc. Inst. Mech. Engrs., 186 (38) (1972).

5. Ciliberti, D. F. and B. W. Lancaster, Chem. Eng. Sci., 31, 409 (1976).

6. Ciliberti, D. F. and B. W. Lancaster, AIChE J., 22, 394 (1976).

7. Ciliberti, D. F. and B. W. Lancaster, AIChE J., 22(6) (1976).

8. Lee, S.H.D. and I. Johnson, 24th Intl. Conf. on Gas Turbines, San Diego (1979).

CILIBERTI, D.F.

RANADIVE, A.Y.

KEAIRNS, D.L.

MULIK, P.R.

ALVIN, M.A.

Desulfurization of Hot Coal-Derived Gas by Calcined Dolomite

Pilot-scale results show that the desulfurization of coal-derived gas at 816 to 982°C for use in a direct-reduction application is feasible.

E.J. Nemeth, J.E. McGreal, J.C. Howell, and J. Feinman
U.S. Steel Corp.
Monroeville, Pa.

Our studies have shown that an efficient coal-gasification-based process for direct reduction of iron ore would include entrainment gasification to make a high concentration of reducing components, a shaft-reduction furnace, and a hot-desulfurization step. To be compatible with the direct-reduction furnace, the desulfurization step should operate in the 816 to 982°C (1500 to 1800°F) range. We have studied the use of packed beds of calcined dolomite to accomplish the gas desulfurization. The calcium component of the dolomite reacts wiuth the H_2S in the coal-gasification gas according to the reaction:

$$CaO + H_2S \rightarrow CaS + H_2O \qquad (1)$$

For a successful process, the dolomite must remove over 90% of the sulfur from the reducing gas and the sulfided dolomite must be returned either to the CaO form or to a form suitable for disposal. Our program considered both the desulfurization of hot-reducing gas by calcined dolomite and the regeneration of the sulfided dolomite at both the bench-scale and pilot-scale level.

Bench-scale gas-desulfurization experiments

Several series of experiments were carried out in a 152-mm (6-in.) deep beds of dolomite held in a heated tubular reactor. In most of the runs, 1.5% H_2S in a hydrogen stream was used. Only particle size exerted a significant effect on the rate or the extent of reaction. For instance 6-mm (¼-in.) by 6-mesh dolomite was saturated with sulfur in 125 minutes, whereas it took 25-mm (1-in.) by 19-mm (¾-in.) dolomite about 275 minutes to reach its final level of reaction, which was 75% of saturation. None of the other variables tested, which included temperatures between 760 and 982°C (1400 and 1800°F), flow rates up to twice normal, water vapor content up to 20%, gas pressures to 1344 kPa (195 psig), and H_2S concentrations up to 3.2%, had a significant effect on the rate or extent of reaction.

Bench-scale experiments on the regeneration of sulfided dolomite

Oxidation reactions to generate SO_2 and CaO from the CaS can have two sources of problems:

1. The possibility of developing excessively high temperatures in the dolomite bed and rendering the stone unreactive, and
2. Conversion of the CaS to $CaSO_4$ rather than to CaO.

To control the heat release during the generation of SO_2 from CaS, we employ the endothermic reaction:

$$CaS + 3\begin{Bmatrix} CO_2 \\ H_2O \end{Bmatrix} \rightarrow CaO + 3\begin{Bmatrix} CO \\ H_2 \end{Bmatrix} + SO_2 \qquad (2)$$

Because of equilibrium limitations in the temperature range of interest, Reaction 2 as written would allow only about 1% SO_2 in the regeneration gas, a level too low for economic recovery. By also introducing O_2 into the regeneration gas, the CO and H_2 reaction products can be consumed thereby shifting Reaction 2 toward the right and favoring an increased yield of SO_2.

Although Reaction 2 is endothermic, the addition of O_2 in concentrations greater than 1% causes the net reaction to become exothermic, although not as exothermic as the reaction of only oxygen with CaS to liberate SO_2. Thus, there is a potential for increasing concentrations of SO_2 in the off-gas as well as controlling temperature rise by using the O_2-CO_2-H_2O system. These gases could be available in the direct-reduction process by combusting spent reducing gas.

Except for runs made on various size fractions of dolomite, the same dolomite was used for 36 consecutive runs in the bench-scale tests with packed beds of dolomite. It was found that:

1. Oxygen concentration of 1024°C (1875°F) inlet gas caused the SO_2 concentration in the product gas to increase from 0.85% SO_2 with no oxygen in the regeneration gas to 2.4% SO_2 with 4.4% O_2 in the regeneration gas.
2. A greater tendency for sulfate formation occurred at higher oxygen concentrations.
3. The concentration of SO_2 in the product gas was higher at higher temperatures.
4. Plus 13-mm (½-in.) sulfided dolomite regenerated slower than smaller sizes.
5. The sulfided dolomite looked like it could be regenerated in cyclic operation; at the end of the 36th regeneration of the same dolomite the residual sulfur was only 0.25%.

Pilot-plant program

The pilot-plant desulfurization vessel consisted of a 254-mm (10-in.) I.D. refractory lined reactor packed with dolomite to a 3-m (10-ft) depth. Reducing gas was made in a natural gas reformer and spiked with 1.5% H_2S. Regeneration of the sulfided dolomite was accomplished with simulated combustion products containing 70% CO_2 and 30% H_2O preheated to 927 to 1038°C

(1700 to 1900°F). Oxygen amounting to 0 to 5% of the total regeneration gas flow was also injected.

In all, the pilot-plant was maintained at operating conditions for 23 days, which included 16 days of actual desulfurization or regeneration and 7 days of various nonprocess operations such as temperature adjustment or maintenance.

Figure 1 shows the effluent gas SO_2 concentration and the temperature response at various levels in the bed for regeneration during cycle 6. The feed-gas O_2 concentration was about 4%, and the inlet-gas temperature and initial bed temperature were about 87°C (1600°F) for cycle 6. These conditions resulted in a temperature response in the bed, with a hot zone peaking between 1121 and 1177°C (2050 and 2150°F). This hot zone followed the reaction-wave front progressing upward through the bed. The SO_2 concentration in the effluent gas peaked at 3.25%. However, this SO_2 concentration was not achieved immediately and the reason appears to be related to the time for the exothermic reactions to heat the bed to a temperature that retards the accumulation of residual calcium sulfate in the bed. The SO_2 concentration tailed off near the end of regeneration when the reactions neared completion.

Desulfurization of the hot-reducing gas was satisfactory in the first part of the campaign. The H_2S concentration in the effluent gas was usually less than 500 ppm. This is equivalent to more than 97% removal of H_2S by the dolomite. The bed lost sulfide-holding capacity during the first seven cycles. The sulfur-removal efficiency of the bed decreased from 83% (initially) to less than 35% by cycle

7. Some of this loss in capacity is due to sulfate formation in the bed, and to the discontinuations of the regeneration step before all the CaS in the bed was converted. Grab samples taken from the bed during operation to verify the presence of both calcium sulfate and sulfide as well as a loss of dolomite surface area in the first seven cycles, Figure 2.

In conclusion

Bench- and pilot-scale results have shown that the desulfur-ization of coal-derived gas at 816 to 982°C (1500 to 1800°F) for use in a direct-reduction application is feasible. For best results, we would use minus-12.7-mm (minus-½-in.) material. Although all our work was done in packed beds, gravitating beds would also appear to be feasible. The use of CO_2, H_2O, and O_2 mixtures for regeneration appears promising in that temperature control is attainable and the SO_2 concentration responds to increased O_2 concentrations as expected. However, it would be desirable to increase the concentration of SO_2 in the product gas, develop a steady evolution of SO_2, and develop means to avoid converting sulfide to sulfate. A decrease in dolomite capacity with cycling, although not as severe as in other methods of regeneration to form H_2S, may also be a problem. The material in this paper is intended for general information only. It should not be used in relation to any specific application without independent examination and verification of its applicability and suitability by professionally qualified personnel. Those making us thereof or relying thereon assume all risk and liability arising from such use or reliance.

McGREAL, J.E. **HOWELL, J.C.** **FEINMAN, J.** **NEMETH, E.J.**

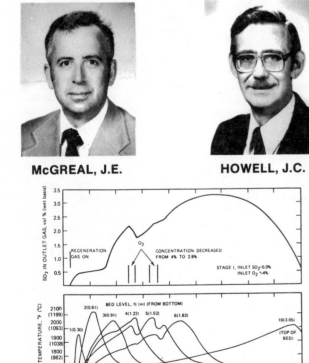

Figure 1. Bed temperature with bed-depth parameters vs. elapsed time and SO₂ content in the outlet gas vs. elapsed time for regeneration cycle No. 6 campaign 1.

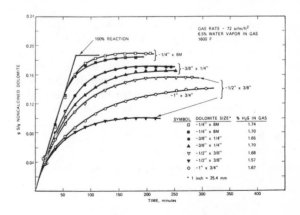

Figure 2. Effect of dolomite size on sulfidation of 152-mm (6-in.) deep bed of moler dolomite.

Gravimelt Process for
Near Complete Chemical Removal

A technology for removal of both 90% of the sulfur and 95% of the mineral matter from coal is discussed.

R.A. Meyers, W.D. Hart, and L.C. McClanathan
TRW Defense and Space Systems Group
Redondo Beach, Ca.

The Environmental Protection Agency has estimated that, even in the year 2000, approximately 75-80% of the total sulfur dioxide emissions in this country will come from power plants and industrial boilers built before (1970) (1).

It is recognized that these existing sources of sulfur oxide air pollution are generally less stringently controlled under the state implementation plans than are the new sources under the Clean Air Act, and, in general, scrubbing of the emissions from these existing plants is neither required nor practicable. Further, fluidized-bed combustion systems cannot help the situation as this would require replacement to the existing boilers.

Coal cleaning or beneficiation is the most promising of all short and near-term control strategies (at least up to the year to 2000). While conventional and improved conventional coal cleaning technologies can reduce emissions sufficiently in less impacted areas, advanced methods will be needed for problem regions.

Chemical coal cleaning could provide the answer for the impacted areas if a process was available that could remove most of the sulfur from coal. Then, chemical desulfurization plants could be built at mine locations and the product could be distributed among a multitude of existing boilers.

The near complete desulfurization and demineralization of coal has been demonstrated via the TRW proprietary Gravimelt Process (2). The sulfur removal step has been further laboratory-demonstrated under the sponsorship of the U.S. Department of Energy (Contract No. DE-AC22-80PC30141). Coal extractions presented in this paper, consistently demonstrate chemical removal of at least 80-90% of the sulfur and 95-96% of the mineral matter from coal. When the sulfur reduction due to cleaning at the mine is added to the total, 90-95% removal is realized, which meets federal pollution control standards as well as state implementation plans for utilization of coal. In addition, the mineral (or ash) content of the coal is reduced by 99%.

The resulting solid coal product is quite unique in that it has no more sulfur or ash than a medium grade fuel oil. In fact, the chemically desulfurized and demineralized coal has sulfur and ash content almost indistinguishable from liquefied coal.

The TRW Gravimelt Process (Figure 1) involves the treatment of mine-cleaned coal with molten potassium and or sodium hydroxide to chemically extract both organic and pyritic sulfur into the molten alkali. The coal mineral content is broken down to forms insoluable in water but soluble in a second liquid[*]. The high density of the melt causes the

---

[*]Proprietary TRW process information

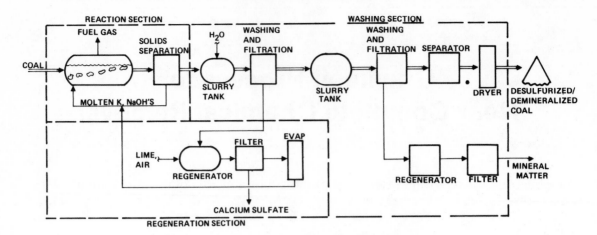

Figure 1. TRW gravimelt process flow sheet.

desulfurized coal to float to the surface, where it is skimmed off. The coal is then washed with water to completely recover the alkali metal and the coal is dried. If the coal is next washed with a second liquid, almost all of the mineral matter is extracted into the solution. Data for both of these steps is presented in the next section.

The use of aqueous caustic to desulfurize coal and the subsequent use of dilute acid to remove mineral matter from caustic treated coal has appeared in the literature. Stambaugh et al (3) reported partial desulfurization and deashing of coal by first reacting the coal with an aqueous solution of mixed caustics under conditions of at least 250°C and 600 psig and then separating and leaching out the hydrolyzed mineral matter with dilute sulfuric acid. This process removed an average of 66% of the coal ash and 66% of the coal sulfur. Similarly, Reggel et al (4) treated coal with aqueous solutions of sodium hydroxides under conditions of at least 225°C and 350 psig, separated the coal and extracted mineral matter with dilute sulfuric or hydrochloric acid to remove up to 93% of the mineral matter but only an average of 25% of the coal sulfur (essentially none of this was organic sulfur) Tippmer et al (5) treated crushed coal first at 20-80 atmospheres and 250-280°C with aqueous sodium carbonate and then washed the coal with water to reduce the ash. No specific ash reductions were given. Masciantonio (6) reported the removal of 80-90% of the sulfur from coal using molten (non-aqueous) caustic under conditions similar to ours but he did not apply a second step to give the essentially hydrocarbon product obtained by the Gravimelt Process.

Thus, our process is unique in two ways 1) both 90% of the sulfur and 90% of the mineral matter is removed giving a coal product which has never before been obtained and which is uniquely useful and 2) both sulfur removal and mineral matter removal are accomplished without the use of high pressure equipment.

LABORATORY RESULTS

A series of experiments were conducted on Kentucky No. 11 seam coal supplied by the Tennessee Valley Authority from its Breckenridge cleaning plant. The feed to the plant contains approximately 4% sulfur and 50% ash and has a heat content of 6800 Btu/lb. The clean coal contains 3-4% sulfur, 7-12% ash and has a heat content of 12500-13500 Btu/lb (dry basis). The sample supplied contained 3.5% sulfur, 7.9% ash and had a heat content of 13180 Btu/lb, thus falling within the normal range for the Breckenridge plant.

An initial series of runs were made with a KOH/NaOH melt (Table 1). The resulting extracted coal contained a mean sulfur content of 0.425 lbs S/10^6 Btu (0.85 lbs SO_2/10^6 Btu) for an average 84% sulfur reduction. There was near 100% removal of pyrite and sulfate sulfur and 70% removal of organic sulfur. The ash content was essentially unchanged by the treatment and there was no significant difference in sulfur removal or ash removal between the two mesh sizes investigated. The MAF heat content values varied from essentially no significant difference (Run 1) to a decrease (Runs 2, 3 and 4).

A pure sodium hydroxide melt was used to desulfurize the Kentucky No. 11 coal

Table 1. Removal of sulfur from Kentucky #11 seam coal by the gravimelt process.[a]

RUN	CONTENT lbs S/10^6 Btu				HEAT CONTENT Btu/lb		ASH % w/w	COAL MESH SIZE
	S_t	S_p	S_s	S_o	Analyzed	MAF		
Control	2.7	1.2	0.0	1.5	13182	14314	7.9	45x200
1	0.4	0.1	0.1	0.3	12812	14124	9.3	45x200
2	0.4	0.0	0.0	0.4	11993	13304	9.8	45x200
3	0.5	0.0	0.0	0.5	12602	13651	7.6	45x200
4	0.4	0.0	0.0	0.4	12495	13646	8.4[b]	-200

[a] Reaction of coal with a 50:50 mixture of KOH and NaOH at 370°C for 20 minutes. All analyses on a dry basis.

[b] No change over input -200 mesh coal ash analysis.

(Table 2). In Runs 5a-c, the same melt was used
sequentially to desulfurize fresh charges of

Table 2. Removal of sulfur and mineral matter from Kentucky #11 seam coal by the gravimelt process.[1]

RUN	CONTENT lbs S/10^6 Btu				HEAT CONTENT Btu/lb		ASH % w/w
	S_t	S_p	S_s	S_o	DRY	MAF	
Control	2.7	1.2	0.0	1.5	13182	14314	7.9
5a	0.6 (0.6)[2]	0.3	0.0	0.3	13565	14666	7.5 (0.3)[2]
5b	0.6 (0.7)[2]	0.2	0.0	0.4	13480	14586	7.6 (0.3)[2]
5c	0.6	0.3	0.0	0.3	13123	14303	8.3
6[3]	0.3	0.2	0.0	0.1	13690	14502	5.6

[1] Reaction of coal with NaOH at 370°C for 20 minutes. All analysis on a dry basis.

[2] Brackets indicate sample was washed with a second liquid after water wash.

[3] Reacted with NaOH melt, washed, dried and reacted again with NaOH melt. Total reaction time was 40 minutes.

control coal to simulate commercial practice of melt recycle. It can be seen that equal desulfurization was achieved for each of the runs, although the melt had been used twice for the desulfurization Run 5c. The mean sulfur content of the three runs was 0.6% with a zero standard deviation which is a 78% reduction in total sulfur content and a 77% reduction in organic sulfur. The mean MAF heat content of the three treated coals was 14518 Btu/lb with a standard deviation of 191 Btu/lb, while the ash mean was 7.8% with a standard deviation of 0.4%. Thus the precision of these three experiments was excellent even though the sulfur compound content of the melt was constantly increasing.

A comparison of the coal heat content results of Tables 1 and 2 indicates that, for this coal, sodium hydroxide extraction has no effect on the MAF heat content of the coal while the addition of potassium hydroxide to the melt apparently causes a decrease in heat content.

A double extraction of the coal with sodium hydroxide was performed (Run 6) for a total retention time of 40 minutes. This resulted in significant additional sulfur removal to a level of 89%. It can be seen that the organic sulfur content of the coal and indeed the pyrite and sulfate content is nearly zero. Again, the MAF heat content is unaffected.

Extraction of the products of Runs 5a and 5b almost totally removed the residual mineral content of the coal giving a solid hydrocarbon fuel similar in sulfur and ash content to a medium grade fuel oil. It should be noted that extraction of coal without prior melt treatment results in only a minor mineral content removal.

The use of potassium and sodium hydroxides are further compared in Table 3 where residual

The same experimental procedures (50:50 mixture of KOH and NaOH) followed by a second wash was applied to samples of three other coal seams: Lower Kittanning (4.05 lbs $S/10^6$ Btu and 13.60% ash); Illinois #6 (2.80 lbs $S/10^6$ Btu and 11.92% ash) and Lucas #5 (1.65 lbs $S/10^6$ Btu and 8.29% ash). The resulting treated coal products contained 0.44 lbs $S/10^6$ Btu and 0.4% ash; 0.2 lbs $S/10^6$ Btu and 0.14% ash; and 0.19 lbs $S/10^6$ Btu and 0.4% ash, respectively. The sulfur removal range for the three coals was 88-93% and 95-99% of the ash was removed. This is quite similar to the results for the Kentucky No. 11.

The coal melt and wash water from Experiment No. 1 of Table 1 were analyzed for sulfur content utilizing the Eschka method for total sulfur as performed by the Warner Laboratories of Cresson, Pa. The results are

Table 3. Removal of sulfur and mineral matter from Kentucky #11 coal and residual alkali metal content.

RUN[1]	MELT	LBS $S/10^6$	HEAT CONTENT		ASH % w/w	RESIDUAL ON COAL % w/w	
			ANALYZED	MAF		Na	K
Control	-	2.7	13182	14314	7.9	.05	0.14
7a	NaOH/KOH	0.57	13517	13517	0.51	.05	0.1
7b	NaOH/KOH	0.56	13361	13383	0.16	.01	.01
8a	NaOH	0.68	14068	14113	0.32	.02	0.004
8b	NaOH	0.58	14023	14055	0.23	.02	.001
9a	KOH	0.24	12897	12965	0.53	.01	.02

[1] Reaction of coal with alkali metal hdyroxide(s) at 370°C for 20 minutes. All analysis on a dry basis. All coal products washed with the second liquid.

coal sodium and potassium content are also measured. It can be seen that the use of sodium hydroxide alone results in slightly less sulfur removal, in a given amount of time, than use of either potassium hydroxide alone or mixtures of sodium and potassium hydroxide. The ash removal is equivalent for all cases but, as in Table 2, the sodium hydroxide Runs 8a and 8b show only small MAF heat content losses (attributable to the replacement of sulfur with oxygen in the coal macromolecule) as compared with potassium hydroxide (Run 9a) or mixtures of potassium and sodium hydroxides (Runs 7a and 7b). All of the residual sodium and potassium levels were equal to or less than the starting alkali metal content of the coal.

presented in Table 4. The 20 gram sample of coal placed in the Gravimelt reactor contained 0.70 grams of sulfur within its organic and inorganic structure (3.5% w/w). The 16.3 grams of coal which were recoved (some coal was lost in handling) contained 0.09 grams of sulfur (0.53% w/w). Thus, 0.61 grams of sulfur had been removed according to Eschka analysis of the "In" and "Out" coal and should be present in either the retrieved melt or wash water.

Eschka analysis of the melt showed that 0.51 grams of sulfur were present, accounting for 80% of the sulfur removed from the coal. Analysis of the first wash water disclosed the remaining sulfur which had been removed

Table 4. Gravimelt process sulfur balance for Kentucky #11 seam coal.[a]

RETRIEVED	IN	OUT
Coal,		
(g)	20.0	16.3
S_t (% w/w)	3.5	0.53
S_t (g)	0.70	0.09
S_t (g) removed		0.61
Melt,		
(g)		176 (173)[b] (181)[c]
S_t (% w/w)		0.29 (0.28)[b] (0.28)[c]
S_t (g)		0.51
First Wash Water,		
(g)		650
S_t (% w/w)		0.02
S_t (g)		0.13
Second Wash Water,		
S_t (g)		0.01
Third Wash Water,		0.01

[a] Example No. 1, Table 2.
[b] Melt from Example No. 2, Table 2.
[c] Melt from Example No. 3, Table 2.

from the coal. The second and third water washes contained near zero sulfur content. Although wash water analyses were not obtained for Runs 2 and 3 of Table 1, the sulfur content of the melt accounts for approximately 80% of the sulfur removed from the coal. The similarities of both the coal sulfur contents and the melt sulfur contents for Runs 1 through 3 again show the precision of the experimentation, sampling and analysis.

Experimentation we performed for the same three runs for determing caustic removal by each of the three water washes (Table 5). It can be seen that most of the caustic (as also most of the sulfur) is removed in the first wash and that the caustic removal becomes vanishingly small for the third wash.

Table 5. Removal of caustic from Kentucky #11 coal by water washing.

RUNS	WASH NUMBER	CAUSTIC CONCENTRATION IN WASH WATER, g/L
1	1	85.8
	2	2.2
	3	0.3
2	1	69.3
	2	1.3
	3	0.1
3	1	89.1
	2	4.4
	3	0.5

REFERENCES

1. Rauch, R.J., Environmental Protection Agency Coal Cleaning Environmental Review, 3 (2), 3 (1980).

2. Meyers, R.A. and W. Hart, Patents Pending.

3. E. P. Stambaugh in Coal Desulfurization, T. D. Wheelock, ed., Am. Chem. Soc. Symposium Series, 64, Washington, D.C. (1977), p.198.

4. L. Reggel, R. Raymond, I. Wender and B.D. Blaustein, Am. Chem. Soc., Fuel Dv. Preprints, 17:(1), 44 (1972).

5. K. Tippmer, H. Schmidt, H. Vinke and G Abenddorth, Australia Patent 505,671 (1977).

6. P.X. Masciantonio, Fuel, 44: 269 (1965).

MEYERS, R.A.

Sulfur Removal from Coal Char Using "Convert-Remove" Technology

With the new EPA sulfur dioxide emission standard which requires 90% removal of all sulfur, the long-term value of partial chemical coal cleaning is questionable. If the option to burn coal cleaned prior to combustion is to exist, improved methods to remove organic sulfur must be found.

A.B. Tipton
Occidental Research Corp.
Irvine, Ca.

BACKGROUND

Occidental's Research Corporation's (ORC) Flash Pyrolysis coal liquefaction process (1) uses rapid heat-up of finely ground coal to produce coal liquids along with a char co-product. Coal chars generally have similar sulfur contents to the parent coal. Hence, high-sulfur, bituminous coals produce high sulfur chars. The value of such char for utility boilers would be increased greatly if its sulfur content met the EPA sulfur emission standard. Hence, concurrent with the development of the Flash Pyrolysis coal liquefaction process, char desulfurization research has been carried out at ORC.

The chemistry of char desulfurization is different from coal desulfurization because the coal pyrite usually has decomposed to pyrrhotite. The conventional approach to char desulfurization is hydrodesulfurization (HDS). HDS uses very high hydrogen to solids ratios to remove sulfur. An example of HDS with char is shown in Figure 1. Here, almost 6000 SCM H_2/ton (200,000 SCF H_2/ton) was required to take 2.4% sulfur to 0.7% sulfur. In this example, even with large volumes of hydrogen, a compliance fuel was not produced. The main chemical reaction which controls the sulfur removal in HDS and demands the high hydrogen capacity is the equilibrium reaction between ferrous sulfide and hydrogen,

$$FeS + H_2 = Fe + H_2S. \qquad \text{Eq. (1)}$$

At 800°C, the concentration of hydrogen sulfide in hydrogen would have to be less than 2000 ppm before this reaction would proceed forward. To keep the hydrogen sulfide dilute enough for sulfur removal to occur requires 3900 SCM of hydrogen per ton of char (125,000 SCF/ton) to remove one percent of sulfur.

Early char desulfurization research at ORC developed an acid leach/hydrodesulfurization process (2) to overcome the equilibrium reaction between ferrous sulfide and hydrogen. The cost of filtering and drying the acid-leached char in this process lead to the search for a totally dry process. These efforts have produced a novel approach for char desulfurization which we call "Convert-Remove" technology.

CHEMISTRY OF THE "CONVERT-REMOVE" TECHNOLOGY

The "Convert-Remove" technology uses two types of treatment steps to produce a low sulfur char. The Convert step is concerned with lowering the organic sulfur while the Remove step only affects inorganic sulfur.

THE CONVERT STEP

The chemistry of the Convert step includes two reactions. First, hydrogen, reacts with organic sulfur (RS) to form hydrogen sulfide,

$$RS + H_2 = R' + H_2S. \qquad Eq. \ (2)$$

Next, the hydrogen sulfide is free to react with in-situ sulfur scavengers to form inorganic sulfides because the hydrogen to solids ratio is low,

$$FeO + H_2S = FeS + H_2O. \qquad Eq. \ (3)$$

The overall result of hydrogen treatment is the conversion of organic sulfur to inorganic sulfur with total sulfur remaining constant. This is in extreme contrast to the conventional approach to hydrogen treatment (HDS) where much higher hydrogen to solids ratios are used and sulfur is removed.

THE REMOVE STEP

The removal of the inorganic sulfide sulfur could be accomplished in a number of ways, e.g., acid leach (2), oxidation. However, we have chosen an approach which regenerates the sulfur scavengers, i.e., steam displacement using the reverse of the reaction by which it was formed,

$$FeS + H_2O = FeO + H_2S. \qquad Eq. \ (4)$$

While high flow rates of steam are now required to sweep out the hydrogen sulfide to maintain removal, very little water is actually consumed - probably no more than a gallon (four liters) per ton of char.

EXPERIMENTAL

Two laboratory batch reactor systems are used for char desulfurization studies. Figure 2 shows the schematic diagram of the reactor setup. One of the reactor systems is equipped with two gas chromotographs for on-line gas analysis. One gas chromatograph (a Perkin Elmer Sigma 1) is used to monitor hydrocarbons and fixed gas composition in steam and hydrogen treatments of solids. The sulfur species such as H_2S, SO_2, COS, CH_3SH, CH_3SCH_3, CH_3SSCH_3 and CS_2 of the gas stream are measured by a Tracor gas chromatograph with a Hall detector.

The high sulfur chars used in the experiments reported in the next section were produced in a bench scale entrained flow reactor using an air-nitrogen mixture as carrier gas. Pulverized West Kentucky No. 9

seam coal (Hamilton Mine) at 635°C (1075°F) for 0.69 sec. with 3% oxygen for decaking was used to produce the coal char.

RESULTS

The strength of the "Convert-Remove" technology comes from the discovery that repetition of short cycles of the two treatment steps is more effective for sulfur removal than is a single two-step cycle with long treatment times. Also we find that when the initial coal devolatilization step has produced a char with a high sulfide sulfur content, an initial Remove treatment step prior to the "Convert-Remove" cycle will produce a lower sulfur product. The results which established these effects are given in Table 1.

In the first series of tests, high volumes of hydrogen were used. In the second series, we reduced the hydrogen volume by an order of magnitude. For an initial Convert treatment we have found that volumes as low as 30 SCM/ton are effective. The results are shown in Figure 3.

The technology has also been tested on a Wyoming sub-bituminous coal and a "Flash Pyrolysis" char (1) from this coal. The results of these tests are shown in Table 2. These data give strong support to our mechanism for organic sulfur removal via an in-situ sulfur scavenger. All treatment times - 5, 10, 15, or 30 min. - with hydrogen produced identical results. The sulfide sulfur capacity of this material is quickly saturated by the hydrogen treatment. Until this sulfide sulfur is removed with steam, the residual organic sulfur is unreactive to the hydrogen. The three step RCR treatment is marginal for ninety percent removal, while the five step RCRCR treatment accomplished almost complete removal.

A side benefit of the "Convert-Remove" technology is the removal of nitrogen. The nitrogen contents of the coals, chars, and desulfurized chars are given in Table 3.

For both the bituminous and sub-bituminous coals about half of the nitrogen was removed.

CHAR REACTIVITY

Reactivity of char to both combustion and desulfurization is important to the utilization of the "Convert-Remove" technology. All of the chars tested in our program were devolatilized in an entrained flow reactor with high heating rates, short residence time and moderate temperature using ORC's "Flash Pyrolysis" technology (1). Such conditions have been shown to be ideal for producing reactive chars (3,4,5,6). Essenhigh (3,4) found that a gasification char had equivalent reactivity to combustion as coal, while a COED char produced with lower heating rates and longer residence times had a much lower reactivity. Walker (5,6) in studying reactivity of chars to gasification, found that both rapid heating and low temperature air oxidation of caking coals enhance the reactivity of the chars produced.

We have found that less sulfur is removed by direct "Convert-Remove" treatment of a decaked high sulfur bituminous coal than is removed from char. However, the same level of sulfur removal is found for sub-bituminous coal and char.

Consequently when caking high sulfur coals are to be desulfurized, an initial coal devolatilization treatment which produces a reactive char, e.g., partial gasification or decaking and rapid pyrolysis, will be needed.

CONCLUSION

The "Convert-Remove" technology is effective in removing ninety percent or more of the sulfur in a reactive char from high sulfur bituminous coals and in sub-bituminous coals or chars.

REFERENCES

1. Che, S.C., Duraiswamy, K., Knell, E.W., and Lee, C.K., "Flash Pyrolysis Coal Liquefaction Process Development," Occidental Research Corporation Final Report to the U.S. Department of Energy, Fe-2244-26, April 1979.

2. Tipton, A.B., "Improved Hydrodesulfurization of Coal Char by Acid Leach," ACS Symposium Series, No. 64, 280 (1977).

3. Essenhigh, R.H. and Csaba, J., Ninth Symposium (Internation) on Combustion, p. 111, Academic Press, 1963.

4. Cogoli, J.G., Gray, D. and Essenhigh, R. H., "Flame Stabilization of Low Volatile Fuels," Combustion Science and Technology, 16, 165, (1977).

5. Ashu, J. T., Nsakala, N.Y., Mahajan, O.P. and Walker Jr., P.L., "Enhancement of Char Reactivity by Rapid Heating of Precursor Coal," Fuel, 57, 250 (1978).

6. Mahajan, O.P., Komatsu, M. and Walker Jr., P.L., "Low Temperature Air Oxidation of Caking Coals, I. Effects on Subsequent Reactivity of Chars Produced," Fuel, 59, 3 (1980).

TIPTON, A.B.

Table 1. Convert (C) and remove (R) treatment steps char desulfurization results.

Process Steps	Total Sulfur	Sulfide Sulfur	Organic Sulfur	SCM H2 Ton	lbSO2 MMBTU
Coal	2.75	0.04	1.74	-	4.2
Starting Char	2.42	0.59	1.73	-	4.2
CR	0.64	0.13	0.43	2,400	1.1
RCR	0.49	0.03	0.37	2,400	0.9
CRCR	0.34	0.04	0.29	4,800	0.6
CRCRCR	0.23	0.05	0.13	7,200	0.4
CCCRRR	0.46	0.18	0.18	7,200	0.8
Starting Char	2.48	0.80	1.59	-	4.3
CR	0.94	0.34	0.52	400	1.6
CCRRR	0.50	0.10	0.27	800	0.7
CRCR	0.43	0.14	0.22	800	0.5
RCRCR	0.27	0.11	0.09	800	0.4

Table 2. "Convert-Remove" treatments of Wyoming sub-bituminous coal and char.

Treatment*	Flash Pyrolysis Char			Coal		
	Total Sulfur	Sulfide Sulfur	Organic Sulfur	Total Sulfur	Sulfide Sulfur	Organic Sulfur
Feed Coal	0.79	0	0.63	0.72	0.02	0.6
Char	0.54	0.06	0.41	-	-	-
C(5)	0.64	0.13	0.49	0.74	0.18	0.48
C(10)	0.65	0.11	0.51	0.68	0.19	0.40
C(15)	0.58	0.15	0.40	0.59	0.19	0.32
C(30)	0.64	0.14	0.47	0.74	0.21	0.4
R(15)	0.36	0	0.34	0.55	0.04	0.4
R(30)	0.35	0	0.32	0.53	0.06	0.4
RCR	0.11	0.03	0.06	0.16	0.04	0.0
RCRCR	0.03	0.01	0	0.07	0.04	0

* C-10 min., R-30 min. unless given in parenthesis

Table 3. "Convert-Remove" treatment also takes out half of the coal nitrogen.

Coal Description	% N (Dry Basis)		
	Feed Coal	Char	Desulfurized Char
W. Ky. No. 9 Seam	1.55	1.65	0.77
Wyoming Sub-bituminous	1.24	1.33	0.64

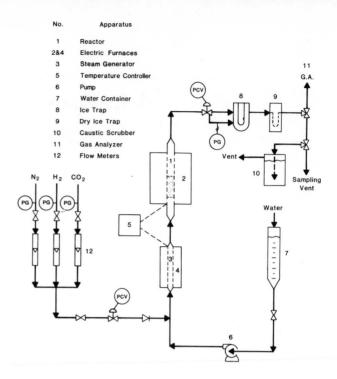

No.	Apparatus
1	Reactor
2&4	Electric Furnaces
3	Steam Generator
5	Temperature Controller
6	Pump
7	Water Container
8	Ice Trap
9	Dry Ice Trap
10	Caustic Scrubber
11	Gas Analyzer
12	Flow Meters

Figure 2. Laboratory batch reactor for "Convert—Remove" treatment.

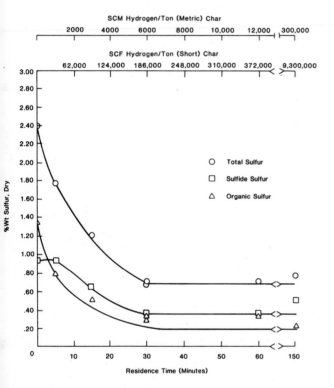

Figure 1. Hydrodesulfurization requires high hydrogen capacity.

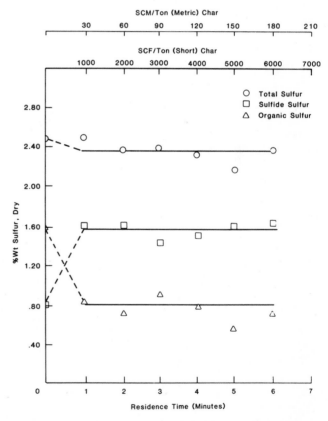

Figure 3. Convert treatment uses low hydrogen volumes.

A Large-Scale Cold Flow Scale-Up Test Facility

This test facility provides opportunity to study the physical phenomena occurring in a commercial-size, fluidized-bed gasifier.

W.C. Yang
Westinghouse R & D Center
Pittsburgh, Pa.

S.S. Kim, and J.A. Rylatt
Westinghouse Synthetic Fuel Div.
Madison, Pa.

A large-scale, cold flow scale-up facility, 3m in diameter by 10m in height and semicircular in cross-section, was constructed with transparent front windows and was commissioned in the latter half of 1980. The test facility provides a unique opportunity to study the physical phenomena occurring in a commercial size fluidized-bed gasifier. The design, objectives, and experimental program are presented. Jet penetration is used as an example to illustrate the utilization of cold flow data for scale-up and to demonstrate the potential usefulness of a large-scale, cold flow test facility as a first step to commercial scale-up.

Because of the abundance of coal reserves in the United States, one of the alternatives for achieving energy self-sufficiency is to convert coal into a clean-burning fuel through coal gasification. Under the sponsorship of the Department of Energy, Westinghouse Electric Corporation is developing an advanced fluidized-bed coal gasification process for low- and medium-Btu gas for power generation and industrial fuel gas. The major emphasis of the program has been the operation of the 15 ton/day Process Development Unit (PDU). A variety of coals, including highly caking eastern bituminous coals, have been successfully fed directly into the gasifier with both air and pure oxygen since 1978 (1).

In support of the PDU operation, cold flow studies have been performed at the Westinghouse R&D Center since 1971 and have contributed significantly to the understanding of the design and operation of the reactor. A large cold flow model test facility, 3m in diameter, has been designed and constructed to study the effect of scale (size) on the important solids flow behavior and gas-solids contacting in the gasifier.

Advantages and Disadvantages of Cold Flow Model Simulation

Cold flow studies have several advantages. Operation at ambient temperature allows construction of the experimental units with transparent plastic material that provides full visibility of the unit during operation. In addition, the experimental unit is much easier to instrument because of operating conditions less severe than those of a hot model. The cold model can also be constructed at a lower cost in a shorter time and requires less manpower to operate. Larger experimental units, closer to commercial size, can thus be constructed at a reasonable cost and within an affordable time frame.

There are limitations on the cold flow model as well. For example, the effects of temperature and pressure are absent. Operating the cold model at atmospheric pressure, nevertheless, will provide useful information and fundamental understanding, as has been effectively demonstrated by past experience in cold flow studies. Since the quality of fluidization and slugging improves with an increase in pressure, data collected in an atmospheric-pressure unit represent, at worst, a conservative case. Adverse phenomena such as temperature gradients, agglomeration, and caking due to temperature effect are more difficult to simulate in an atmospheric cold model, but studies in an atmospheric cold model of coal feed point, solid withdrawal point, stagnation, and solid mixing in the bed will help to avert those problems.

Review of Earlier Cold Flow Simulation Studies

Specific examples from past cold flow simulation studies, where the data collected from the cold model and the mathematical correlations developed based on those data were found to be directly applicable to the PDU conditions, can be grouped into the following categories:

- For quantitative extrapolation - prediction of slugging bed height (2) and jet penetration depth (3).

- For qualitative evaluation - selection of distributor plate designs (4) and the effect of expanded section on slugging (5).

- For predicting trends - effect of jet velocity on solids entrainment into the jet and on solid circulation rates (6,7).

- For feasibility studies - evaluation of the multiple air tube concept (8), of the largest operable air tube, and of char-ash separation (9,10,11).

- For identifying critical design problems and design parameters - identification of the optimum air-tube location (6) and downcomer/draft-tube area ratio (7).

- For design and scale-up - mathematical model developed for scale-up of the devolatilizer (7) and design considerations for commercial fluidized-bed agglomerating combustor/gasifier (12).

Other areas, where the support rendered by the cold flow studies was significant, are personnel and operator training, providing troubleshooting capability (13,14), supporting the PDU operation, and studying the different design and operating parameters.

OBJECTIVES OF THE COLD FLOW SCALE-UP TEST FACILITY

Cold flow model studies in a large-scale unit are an integral part of the Westinghouse scale-up strategy. The strategy, limitations, and needs have been discussed previously (15). The structure of the Westinghouse scale-up strategy can be summarized in Figure 1. The figure depicts the progression from exploratory studies through bench-scale and pilot-plant studies to eventual commercialization. Cold flow models were employed successfully in both bench-scale and pilot-plant studies to gain fundamental understanding of the critical phenomena occurring in the reactor, to generate hydrodynamic models for scale-up, and to

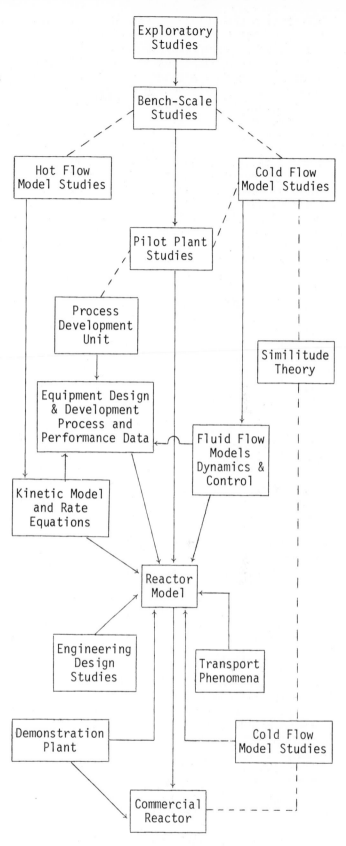

Figure 1. Structure of reactor scale-up.

investigate the start-up, shutdown, and turn-down strategies. The results of cold flow model studies were then combined with kinetic models and rate equations generated from the bench-scale hot models and the data obtained from the pilot-scale PDU to construct a realistic mathematical model for predicting reactor performance.

To ensure successful scale-up, the hydrodynamic models and similarity criteria developed from small units to describe different critical phenomena will have to be verified or modified using data obtained in a large-scale unit. Since it is both time-consuming and prohibitive in cost to construct a full-scale commercial reactor for generating the desired data, a large-scale cold flow model approaching the size of the commercial reactor will be cost effective in the overall scale-up strategy. The effects of pressure and temperature, which are absent in the cold flow studies, can be obtained by analyzing and comparing the data from the PDU and the pilot-scale cold model. They can also be deduced from the data published in the literature. An example on jet penetration will be given later to elucidate the effectiveness of this approach.

The 3-m cold flow model, designed for easy visibility, accessibility, and flexibility in changing into different gasifier configurations, will provide a unique opportunity to study the physical phenomena occurring in a commercial-size plant. Areas where substantial improvement in understanding the gas and solids flow behavior will result are:

- Commercial-size design limitations. The largest single oxidant tube employable commercially without an adverse effect on reactor performance can be identified. If multiple tubes are necessary, the jet interaction and operational stability can be studied and the optimal tube arrangement can be identified.

- Gases and solids distribution in the commercial size equipment. Because of the size of the commercial plant, distribution of gases and solids can become a critical design problem. These include the conical grid in the gasifier and the gas distributor for the stripping gas in the char-ash separator. Stagnation regions are also more probable in a large unit if the gas distribution grids are not carefully designed.

- Char-ash separator design. Clean char-ash separation is important in overall efficiency of the gasification process. The

3-m model offers a unique opportunity for studying the operating conditions and design parameters to achieve the optimal separation in commercial size equipment.

- Standpipe and pneumatic transport line design. Standpipe and pneumatic transport lines can provide data and experience in transferring and circulating a large quantity of solids from cyclones or other reactors.

- Start-up, shutdown, and turndown. Start-up, shutdown, and turndown are critical subject to be studied in a commercial plant. The 3-m model with the capability of continuous feeding and withdrawing bed material can fill this unique need.

- Maximum bubble size. Researchers disagree about whether a maximum bubble size exists in a fluidized bed. If the maximum bubble size exists, a big bubble will split into smaller ones in the fluidized bed so that none of the bubbles is greater than the maximum size. This maximum size may, in turn, dictate the number and size of the oxidant tube(s) feasible in a commercial size plant.

Thus, the primary objectives of the cold flow scale-up test facility are to study the size effect on different critical phenomena occurring in a Westinghouse gasifier and to verify the similarity criteria developed from small cold flow units. The secondary objectives are to use the test facility for trouble shooting and for operator training.

DESIGN OF COLD FLOW SCALE-UP FACILITY

The cold flow scale-up facility located at Waltz Mill, Pennsylvania is shown in Figures 2 and 3. The test facility has the following design features:

- The vessel is semicircular in cross-section and 3m in diameter, with transparent Plexiglas plates at the front (flat side) and transparent windows at the back (circumferential side).

- Bed depths up to 6.1m, gas velocities up to 4.1 m/s and particle sizes up to 0.64 cm can be employed.

- Provisions are provided to test 1, 4, or 7 oxidant-nozzle designs.

- One or two solids feed streams can be provided via two solids feed hoppers.

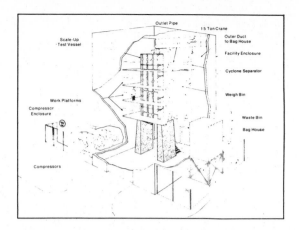

Figure 2. Schematic of the 3-m cold flow test facility.

- Pneumatic transport lines up to 20 cm in diameter are provided.

- Two solids feed options are provided.

- There is also provision for continuous solid fines recycle back into the bed.

- The model can be operated continuously with solids feed and withdrawal.

- Standpipe operation is considered for char recycle line.

A simplified flow sheet is shown in Figure 4. When the bed materials arrive on site, they are unloaded and transported into the storage hoppers (D-M021 to D-M023) by a loading bin and then into the feed hoppers (C-M003 and C-M004) pneumatically. In these feed hoppers, the solids are weighed and fed into the fluidized bed (C-M001) at controlled rates via rotary feeders.

Exhaust gas from the reactor is treated in the cyclone, C-M002. The solids collected in the cyclone flow down in the dip leg and are recycled to the reactor by a radial or an axial feed line. The gas from the cyclone will be vented into the baghouse (T-M031).

The solids withdrawn from the reactor can

Figure 3. A pictorial view of the 3-m cold flow test facility.

be recovered from the recovery cyclone, C-M006. The discharge from C-M006 can be fed to one of the three storage hoppers, to either one of the two feed hoppers, or to the waste bin. By feeding the discharge from C-M006 to one of the feed hoppers, a continuous operation with simultaneous solids feed into and withdrawal from the reactor can be performed at a steady state for an indefinite time.

The bed material to be used initially will be -10 +60 mesh crushed acrylic plastic particles with an average size of 1000 μm and a density of 68 lb/ft^3. The crushed acrylic plastic particles have a shape and density comparable to those of the char particles.

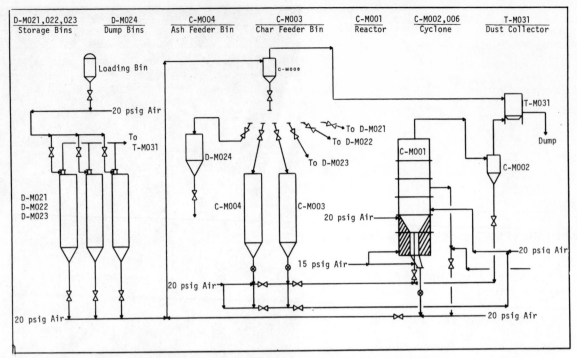

D-M021,022,023	D-M024	C-M004	C-M003	C-M001	C-M002,006	T-M031
Storage Bins	Dump Bins	Ash Feeder Bin	Char Feeder Bin	Reactor	Cyclone	Dust Collector

Figure 4. Simplified scale-up test facility flow diagram.

For simulation of char-ash separation, three pairs of materials of different characteristics in terms of density, size, and shape will be employed.

The test vessel is comprised of six identical, semicircular sections (Figure 5). Viewing from the front face is provided by transparent panels. On the curved face viewing is limited to individual circular windows whose location and number are dependent on the expected fluidized-bed location. Attached to the underside of the bottom viewing section is an ash/char draw-off cone with a feed tube assembly. The draw-off cone fixes the feed tube laterally within the model and provides for vertical adjustment of the feed tube relative to the steam grid cone.

The steam grid cone is located internally within the lower viewing sections. It forms a conical annulus around the feed tube. The conical surface is perforated and is subdivided into several annular regions, any or all of which may be supplied with pressurized air. Bed sample tubes are positioned within the test section for direct withdrawal of materials from within the interior of the bed. The design of the model and process piping provide features

that facilitate changes in feed tube size and configuration, draw-off cone configuration, and steam grid cone.

EXPERIMENTAL PROGRAM

The experimental program will concentrate initially on operating the 3-m model in the gasifier configuration to support the scale-up study of the Westinghouse single-stage gasification system. Data will be collected on the pneumatic transport lines during transport of solids between the storage hoppers and the feed hoppers and also during the simulation of pneumatic transport feed of solids materials into the reactor. The standpipe flows in the char draw-off pipe and in the dipleg of the main cyclone will also be studied.

Experimental Program on Gasifier Configuration

The objective of this program is to study the limitation of physical size of a single air (or oxygen) tube employable in a large-size reactor in terms of operability of the reactor. If multiple tubes are indeed necessary, their design, operability, and jet interaction will be studied. Simulation of char-ash separation will also be carried out in char-ash separators

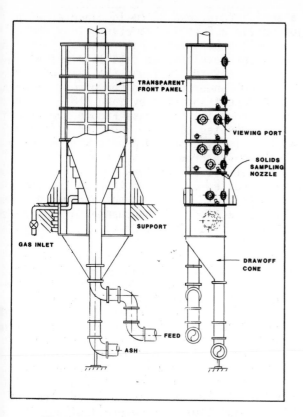

Figure 5. Internal design features.

Table 1a   Design Parameters for the Gasifier

Design Parameters	Phenomena to be Observed
Air tube size and number	Jet expansion and penetration
	Bubble size
	Solid circulation rate
	Interaction between jets for multiple tube configuration
Air tube location	Solid circulation rate
	Stagnation
Conical grid angle	Solid circulation
	Stagnation
Char-ash separator	Char-ash separation
	Operability
	Slugging

f different heights and diameters to investigate the design possibilities and performance. Existing correlations on jet penetration depth, s, and solids entrainment into the jet, and article separation will be compared with the ata obtained from the 3-m cold model to evaluate the possibility of using the existing correlations for scale-up purpose.

The design and operating parameters to be studied and the phenomena to be investigated re summarized in Table 1a and 1b.

The experiments with multiple air tubes ill be similar to those with a single air ube. In addition, jet interaction between eighboring air tubes and bubble coalescence ll be investigated by changing the spacing etween the air tubes.

Start-up, shutdown, turndown, and potentially unbalanced operation among the air tubes ll be studied. Unbalanced operation, where r tubes are not operating at the same velo ty, will be purposely created to study the nge of operability. This type of operation n be later incorporated into the overall rndown strategy.

Table 1b   Operating Parameters for the Gasifier

Operating Parameters	Phenomena to be Observed
Jet velocity	Jet penetration depth
	Solid circulation rate
	Gas and solid entrainment
	Bubble size
Conical aeration flow	Solid circulation
	Stagnation
Solids loading in feed line and velocity	Jet penetration depth
	Solid circulation
	Gas and solid entrainment
Aeration in char-ash annulus	Char-ash separation
Ash withdrawal rate	Char-ash separation
	Char-ash interface location
Aeration location on the conical grid	Solid circulation
	Stagnation
	Gas and solid entrainment into jet

ANALYSIS OF COLD FLOW DATA FOR SCALE-UP — JET PENETRATION AS AN EXAMPLE

The gases are fed into the Westinghouse gasifier through either multiple grid jets or a single large jet. The solids are fed via pneumatic transport lines, and they appear as jets in the reactor as well. Hence, the importance of jetting phenomena in the reactor design is evident. The data available in the literature are meager (6). The largest jet nozzle experimentally studied was only about 6 cm in diameter. Even the definition of a jet and its penetration depth are ambiguous and uncertain (16). Bubbles, pulsating jets, and permanent jets have all been observed above jet nozzles by different authors employing different bed materials and bed configurations and under different operating conditions. More experimental data and theoretical studies will be required to describe these different flow regimes quantitatively. To predict jetting phenomena for jet nozzles up to 80 cm in diameter, the size of oxidant tubes projected for commercial plants, experimental studies in a large-scale cold model are necessary.

Jetting phenomena in a Westinghouse gasifier were first studied in a semicircular column using hollow epoxy spheres (ρ_p = 210 kg/cm^3) as the bed material and air as the fluidizing medium to simulate the particle/gas density ratio in an actual gasifier operating at 1520 kPa and 1280 K (6). A two-phase Froude number defined as $\rho_f U_0^2/(\rho_p - \rho_f)gd_0$ and derived from both the momentum balance and the dimensionless analysis was found to correlate the data well. The same two-phase Froude number was later used to correlate the literature data of multiple vertical grid jets successfully (16). The experimental data included in the correlation, however, are from jets of less than 6 cm in diameter, and the operating conditions are usually at ambient temperature and pressure. The question often asked is whether the developed correlation is applicable for scale-up to a much larger jet nozzle operating at much higher temperature and pressure. The size effect will have to be verified with operational data from a jet nozzle considerably larger than the ones now being employed. The effects of pressure and temperature, however, can be obtained from small-scale units.

A recent study by Knowlton and Hirsan (17) reported the effect of pressure on jet penetration up to a pressure of 5300 kPa (~53 atm) for fluidized beds of sand (ρ_p = 2629 kg/m^3), FMC char (ρ_p = 1158 kg/m^3), and siderite (ρ_p = 3988 kg/m^3). The experiments were performed in a semicircular bed 30 cm in diameter with a jet nozzle 2.5 cm in diameter. They compared their experimental data with the existing correlations developed on data obtained at atmospheric pressure and found that all of the correlations underpredicted the effect of pressure. A recent analysis by Yang (18), however indicated that the two-phase Froude number originally suggested could be modified slightly to account for the pressure effect. The original two-phase Froude number took into account the effect of pressure on the dynamic pressure created by the jet through the momentum flux term $\rho_f U_0$ but did not include a similar correction for the changes of fluidized bed properties due to pressure.

It was found out experimentally that the voidage at minimum fluidization was independent of pressure (19), though the minimum fluidization velocity did change with the operating pressure. Hence, the effect of pressure on the jet penetration depth is not due to changes in voidage but due to changes in viscous and inertial forces of fluid medium on solid particles in the bed. The effect of viscous and inertial forces of fluid on solid particles in a fluidized bed can be calculated from the Ergun equation (20). The overall effect is manifested through the changes of minimum fluidization velocity due to changes in pressure. Thus, it is natural to include the minimum fluidization velocity as a correlating parameter as well. For bed material of wide size distribution, the minimum fluidization velocity determined through the conventional procedure can no longer be used (21). The complete fluidization velocity, the velocity at which the bed becomes completely fluidized, should be employed.

If the data by Knowlton and Hirsan (17) are correlated with a modified two-phase Froude number, a successful correlation results, as shown in Figure 6 and Equation (1).

$$\frac{L_{max}}{d_0} = 7.65 \left[\frac{1}{R_{cf}} \cdot \frac{\rho_f}{(\rho_p - \rho_f)} \cdot \frac{U_0^2}{gd_0} \right]^{0.472} \tag{1}$$

where

$$R_{cf} = (U_{cf})_p / (U_{cf})_{atm}$$

The jet penetration data are successfully correlated to ±40% for sand, FMC char, and siderite up to a pressure of 5300 kPa (~53 atm). It is even more important to note that the limiting form of Equation (1) at 101.4 kPa, where the correction factor R_{cf} = 1, approaches

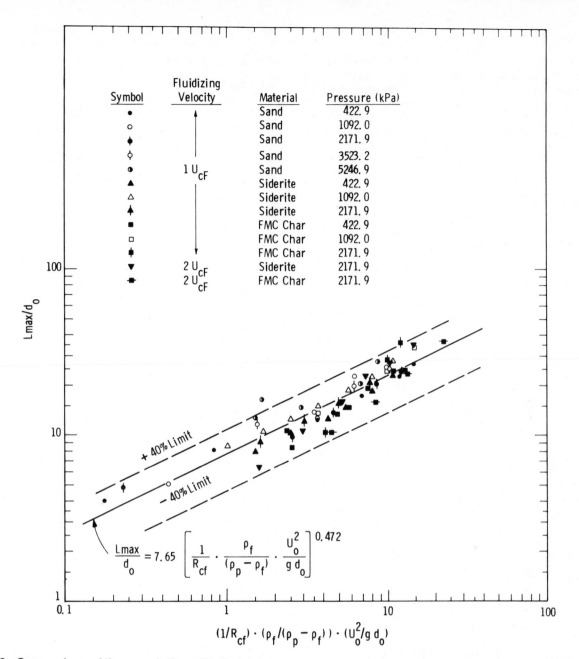

Figure 6. Comparison of the correlation with the high-pressure jet penetration data of Knowlton and Hirsan(17)

the correlation proposed earlier for atmospheric condition, as shown in Equation (2) (6).

$$\frac{L_{max}}{d_o} = 6.5\left[\frac{\rho_f}{(\rho_p - \rho_f)} \cdot \frac{U_o^2}{gd_o}\right]^{0.5} \quad (2)$$

Equation (2) now is only a special case of the general correlation expressed in Equation (1). The good agreement obtained for all data using the modified Froude number signifies the physical significance of the parameter.

The same approach developed here can be employed to build in the effect of temperature. Unfortunately, a literature survey did not come up with usable data. A small-scale, high-temperature test facility will be constructed to study the effect of temperature on jet penetration.

The gas velocity profiles inside the jet region were also investigated for both pure gas jets and gas-solids two-phase jets (22,23). Tollmien's similarity, which is applicable for turbulent jets in a homogeneous medium, was found to approximate the velocity profiles of jets in a fluidized bed as well. This finding is essentially similar to that observed by Massimilla et al. (24) and Donsi et al. (25), as well. The identification of Tollmien's similarity in velocity profiles is significant. The Tollmien similarity can now be employed to estimate the gas entrainment into jets in fluidized beds. The qualitative and quantitative observations in cold flow models on jet penetration, jet expansion, and gas and solids entrainment into the jet can be incorporated in mathematical modeling to develop a more realistic physical model of jetting phenomena in fluidized beds. A recent independent study by Blake et al. (26) on computer modeling of coal gasification reactors indicated that the proposed jetting correlations derived from cold flow models agree fairly well with their numerical calculations based on the coupled heterogenous/homogeneous chemistry and hydro-dynamics of solid particle and gas flows within the coal gasifier at high-temperature and high pressure.

CONCLUSION

An integrated scale-up strategy, of which the cold flow model study is an integral part, has been presented. The usefulness of the cold flow model data in increasing the fundamental understanding of the critical phenomena occurring in the Westinghouse gasifier, in helping to interpret the data obtained from the pilot-scale Process Development Unit, and in developing scale-up criteria was elucidated. Jet penetration was used as an example to illustrate the utilization of cold flow data for scale-up and to demonstrate the potential of a large-scale cold flow test facility as a cost-effective first step to commercial scale-up.

Similar approaches are being pursued to develop fundamental understanding of other critical phenomena occurring in the process, notably the ash agglomeration mechanism, char-ash particle separation, fines recycle, start-up, turndown, and shutdown. A commercial size cold flow test facility will provide further insights into the design and operation of a commercial gasifier.

ACKNOWLEDGMENT

This work was performed under DOE Contract EF-77-C-01-1514. The contract monitors Messrs. M. Carrington and D. Cicero provided valuable suggestions during the course of this program. Direction provided by Drs. D. H. Archer and D. L. Keairns at the R&D Center and Messrs. L. A. Salvador, P. Cherish, and D. Marburger at the Synthetic Fuel Division was instrumental in successfully carrying out this program.
Messrs. R. S. Mavretish, J. E. Macko, R. J. Faix, and others contributed to the design of the test facility.

NOTATION

d_o	jet nozzle diameter
d_p	particle size
g	gravitational acceleration
R_{cf}	ratio of complete fluidization velocity at pressure P over that at atmospheric pressure, $(U_{cf})_p/(U_{cf})_{atm}$
U_{cf}	complete fluidization velocity
$(U_{cf})_{atm}$	complete fluidization velocity at atmospheric pressure
$(U_{cf})_p$	complete fluidization velocity at pressure P
U_o	average jet nozzle velocity
ρ_f	fluid density
ρ_p	particle density

LITERATURE CITED

1. Salvador, L. A., P. Cherish and E. J. Chelen, "Operating experience with the Westinghouse gasifier," presented at the 88th Nat'l. Meeting, AIChE, Philadelphia (1980).

2. Yang, W.-C. and D. L. Keairns, IEC Process Des. Dev., 17 (3), 215 (1978).

3. Yang, W.-C. and G. B. Haldipur, "Performance of a pilot-scale ash agglomerating combustor/gasifier," presented at the 71st Annual Meeting, AIChE, Miami Beach (1978).

4. Yang, W.-C., E. J. Vidt and D. L. Keairns, U.S. patent 4,035,152 to Westinghouse Electric Corp.

5. Yang, W.-C. and D. L. Keairns, AIChE J., 26 (1), 144 (1980).

6. Yang, W.-C. and D. L. Keairns, "Design and operating parameters for a fluidized bed agglomerating combustor/gasifier," Proceedings of the Second Engineering Foundation Conference on Fluidization, p. 208, Cambridge University Press, Cambridge (1978).

7. Yang, W.-C. and D. L. Keairns, AIChE Symp. Series No. 176, Vol. 74, 218 (1978).

8. Salvador, L. A. and D. L. Keairns, "Advanced coal gasification system for electric power generation," Quarterly Progress Report for the period October 1 to December 31, 1978, prepared for U.S. DOE.

9. Chen, J.L-P. and D. L. Keairns, Can. J. Chem. Eng. 53, 395 (1975).

10. Chen, J.L-P. and D. L. Keairns, IEC Process Des. Dev. 17, 135 (1978).

11. Yang, W.-C. and D. L. Keairns, "Rate of particle separation in a fluidized bed," presented at the 73rd Annual Meeting, AIChE, Chicago (1980).

12. Merry, J. M. D., J.L-P. Chen and D. L. Keairns, "Design considerations for development of commercial fluidized bed agglomerating combustor/gasifier," Fluidization Technology, p. 423, Hemisphere Publishing Co., Washington (1976).

13. Yang, W.-C., P. J. Margaritis and D. L. Keairns, Coal Processing Technology 3, 166 (1977).

14. Yang, W.-C., P. J. Margaritis and D. L. Keairns, "Simulation and modeling of start-up and shutdown in a pilot-scale recirculating fluidized bed coal devolatilizer," AIChE Symp. Series No. 176, Vol. 74, 87 (1978).

15. Keairns, D. L., "Fluidization large scale facilities and scale up: strategy, limitations, needs," presented at the National Science Foundation Workshop on Fluidization and Fluid-Particle Systems - Research Needs and Priorities, Rensselaer Polytechnic Institute, Troy (1979).

16. Yang, W.-C. and D. L. Keairns, I&EC Fundamentals 18, 317 (1979).

17. Knowlton, T. M. and I. Hirsan, "The effect of pressure on jet penetration in semicylindrical gas-fluidized beds," Proceedings of the Third Engineering Foundation Conference on Fluidization, Henniker (1980).

18. Yang, W.-C., "Jet penetration in a pressurized fluidized bed," paper submitted for publication in I&EC Fundamentals (1980).

19. King, D. F., "Fluidization under pressure," Ph.D. dissertation, University of Cambridge (1979).

20. Ergun, S., Chem. Eng. Prog. 48 (2), 89 (1952).

21. Knowlton, T. M., "High-pressure fluidization characteristics of several particulate solids, primarily coal and coal-derived materials," AIChE Symp. Series No. 161, Vol. 73, 22 (1977).

22. Yang, W.-C. and D. L. Keairns, "Momentum dissipation of and gas entrainment into a gas jet in a fluidized bed," presented at the 72nd Annual Meeting, AIChE, San Francisco (1979).

23. Yang, W.-C. and D. L. Keairns, "Momentum dissipation of and gas entrainment into a gas-solid two-phase jet in a fluidized bed," Proceedings of the Third Engineering Foundation Conference on Fluidization, Henniker (1980).

24. Massimilla, L., G. Donsi and N. Migliaccio, "The dispersion of gas jets in two-dimensional, fluidized beds of coarse solids," presented at the 72nd Annual Meeting, AIChE, San Francisco (1979).

25. Donsi, G., L. Massimilla and L. Colantuoni, "The dispersion of axisymmetric gas jets in fluidized beds," Proceedings of the Third Engineering Foundation Conference on Fluidization, Henniker (1980).

26. Blake, T. R., D. H. Brownell, Jr., P. J. Chen, J. L. Cook and G. P. Schneyen, "Computer modeling of coal gasification reactors," Annual Report (Year 4), submitted to DOE by Systems, Science and Software (1980).

YANG, W.C.   **KIM, S.S.**   **RYLATT, J.A.**

Desulfurization of Iron Disulfide: Evaluation of Alternative Mechanisms

This research suggests increased surface area may be at least as important and perhaps the dominant factor in the higher reactivities.

D.F. Daugherty
Research Triangle Institute
Research Triangle Park, N.C.

IRON DISULFIDE IN COAL

Form

The total sulfur content of coal varies from less than 0.5 to over 10 weight percent. Most coals in the United States contain between 1 and 3 percent total sulfur with usually 1/2 to 2/3 of the total sulfur being pyritic, and essentially all of the remainder being organic sulfur (1). (Levels of sulfate sulfur greater than 0.1 weight percent almost certainly indicate coals which have been subject to weathering since their initial formation (2).) Pyritic and organic sulfur are thus the two main forms of sulfur in coal whose behavior is important.

Pyritic sulfur occurs as iron disulfide, FeS_2. There are two naturally occurring forms of FeS_2 which differ in their crystalline structure--pyrite which has a cubic crystal structure, and marcasite which has a rhombic crystal structure. Most of the pyritic sulfur in coal has the pyrite structure, and since the chemical reactivity of the two forms is similar, their reactions are rarely considered separately.

King (3) has reviewed several of the classification schemes (4, 5) for iron disulfides present in coal, and examined a West Virginia Waynesburg coal using a classification based on five groups:

1. Framboids -- Derived from French for raspberry, the term framboid denotes a spherical aggregate or cluster of pyrite microcrystals. The microcrystals are uniform in size in a given framboid, but there is no apparent relation between framboid size and microcrystal size. The Waynesburg coal had framboids 2 to 100 micrometers in diameter, and the microcrystals were 0.1 to 2.0 micrometers in diameter.

2. Euhedral Aggregates -- Euhedral refers to crystals which have developed their own shape and crystal faces instead of having their shape impressed by adjacent crystals. Euhedral aggregates are groupings of euhedral microcrystals in contact with one another. They are closely related to and may contain framboids.

3. Isolated Euhedra -- These are isolated occurrences which display distinct crystal faces. They ranged from one to several hundred micrometers in Waynesburg coal.

4. Fracture Fillings -- Euhedral pyrite crystals commonly are found as filling in fractures in the coal seam. The fillings may be from one micrometer to several millimeters in width and up to tens of centimeters long.

5. Irregular Shapes -- The majority of iron disulfide in Waynesburg coal was in irregularly shaped occurrences. The occurrences varied from dendritic, finger-like shapes to smooth spheres. In some of the irregular particles, the remains of the plant cellular structure could be recognized.

While pyrite in coal may occur as large nodules over one meter in diameter, the majority on a mass basis is present as finely divided and well disseminated particles (6). Workers at Penn State (7) have used a technique called automatic reflectance microscopy (ARM) to automatically measure the size distribution of pyrite occurring in coal. The technique measures chord lengths of pyrite particles detected by optical means and automatically prepares a size distribution. From the size distribution, the total amount of pyrite can be estimated. The median chord size varied between 10 and 40 micrometers for several coal samples (8). Note that the pyrite particles are mainly smaller than finely pulverized coal (minus 200 mesh coal is less than 74 micrometer in diameter).

Reactions

Iron disulfide (FeS_2) decomposes to produce gaseous sulfur compounds upon heating. The product gas and rate of decomposition depend upon the atmosphere surrounding the FeS_2. Thompson and Tilling (9) studied the decomposition of iron pyrite in several atmospheres. A sample boat containing pyrite was slid into a furnace preheated to constant temperature. After two hours, the heat was removed and the sample was allowed to cool under the atmosphere being studied. The solid was analyzed for sulfur to obtain conversion.

In carbon dioxide, desulfurization commenced at about 580°C, and at 670°C the sulfur in the solid residue was 35.7 percent, which corresponded closely to FeS. Between 670°C and 900°C, there was no further decomposition of the solid. Then at temperatures above 900°C, the solid began to lose more sulfur. It was speculated that temperatures above 900°C dissociates the otherwise inert carbon dioxide into carbon monoxide and oxygen which then react with the FeS. (Results for decomposition in nitrogen obtained in the current study show some desulfurization of FeS at temperatures around 1000°C. This argues against Thompson and Tilling's proposal of gaseous dissociation being responsible for the second stage of decomposition since nitrogen would not be expected to dissociate.)

In a carbon monoxide atmosphere, desulfurization began at low temperatures, 359°C, and was explained by the reaction:

$$FeS_2 + CO \rightarrow FeS + COS.$$

At about 590°C, the composition corresponded to FeS. Between 590 and 800°C, there was very little further decomposition. In runs above 800°C, Thompson and Tilling encountered anomalous results: there was an increase in the sulfur remaining in the solid. This was attributed to partial fusion of sulfide on the sample surface which then hindered further sulfur removal. (At the slow heating rates used in the current study, no such hindrance was seen.)

Hydrogen commenced removal of sulfur pyrite at about 400°C via the proposed reaction $FeS_2 + H_2 \rightarrow FeS + H_2S$. Desulfurization to FeS was essentially complete at 500°C, and the FeS was stable to about 900°C. Above 900°C, slow desulfurization of FeS was observed, and was explained by reduction to metallic iron, $FeS + H_2 \rightarrow Fe + H_2S$.

There is a common thread running through the results observed by Thompson and Tilling for desulfurization of FeS_2. Iron disulfide loses one sulfur atom at a temperature of 300°C to 600°C to form FeS. The FeS is relatively stable at least until temperatures approach 900°C, at which point it may begin to decompose. For the most part, results from the current study are similar to those of Thompson and Tilling, although the second stage of desulfurization in hydrogen was found to begin at lower temperatures.

BACKGROUND OF THIS STUDY

As part of an EPA-sponsored investigation into pollutant generation during coal gasification, Research Triangle Institute (RTI) has studied the decomposition of FeS_2 under reducing conditions. The technique of nonisothermal reaction analysis (or more simply, thermal analysis) was used to continuously monitor the progress of reactions in atmospheres of nitrogen, hydrogen, and carbon monoxide.

Conventionally, reaction rates are measured by monitoring the extent of reaction in a series of constant temperature runs. The progress of reaction yields the rate of conversion at each temperature. Next, in a separate step, the temperature dependence of the reaction rate is established. This approach simplifies the analysis of data by isolating temperature and composition effects, but is nonapplicable for some reactions. In particular, reactions which occur over a broad temperature range cannot effectively be studied by isothermal techniques--a significant portion of the reaction occurs during the heatup period.

There is another approach known as nonisothermal reaction analysis, or more commonly, thermal analysis. In this technique, the temperature dependence of the reaction rate is recognized and included in the mathematical analysis of experimental data. By closely controlling the temperature program (usually a linear heating rate is used), the same kinetic parameters can be extracted from nonisothermal data as can be extracted from a series of isothermal experiments. In addition, the nonisothermal technique can resolve parallel or sequential reactions which cannot be recognized as separate events in isothermal experiments (10).

The key to successful kinetic studies by thermal analysis lies in relating the extent of reaction to some continuously measurable property of the sample. Two properties most often measured are the heat release during reaction or the sample weight.

In the current study sample weight was monitored. For the gas-solid reactions undergone by FeS_2, a weight loss of 26.7 percent corresponds to the loss of one sulfur atom and if both sulfur atoms are lost,

53.4 percent of this initial weight is lost. If the decomposition reactions occur in two distinct steps, two plateaus of constant weight are observed on the plot of sample weight versus temperature.

Figure 1 displays the nature of data obtained from a nonisothermal experiment. A sample of mineral FeS_2 from Mexico was ground and screened to 16 x 28 Tyler mesh (589 x 991 micrometer) and heated in hydrogen at 8°C/min. Percent sample weight is plotted versus sample temperature. (Although the data were taken as a function of time, runs are more easily compared when plotted versus temperature. The rate of weight loss is also presented on Figure 1. Such derivative plots permit subtle changes in reaction rates to be more easily distinguished.

The two large peaks are shown on the Figure 1 derivative curve. They have the characteristic shape of nonisothermal reaction data--the initial portion of the curve rises steeply due to exponential dependence of reaction rate on temperature. However, the curve cannot rise indefinitely because the rate also depends on the amount of sample remaining. Hence, a maximum rate of weight loss is obtained when the effect of higher temperature becomes overwhelmed by the effect of decreasing reactant. The temperature at which this maximum occurs is the most widely used descriptor of nonisothermal data. For a simple first-order reaction model, the kinetic parameters can be derived from the maximum temperatures obtained at varying heating rates (11). Other techniques exist for extracting kinetic parameters for simple reaction models and several have been critically reviewed in a paper by Flynn and Wall (12).

MODEL SELECTION

The failure of nonisothermal, or for that matter, isothermal rate investigations is usually not in the technique chosen to reduce the data. The more important step is selection of an appropriate reaction model. Ozawa (10, 13, 14) showed how a given set of kinetic data may fit more than one reaction model and result in widely varying kinetic parameters. Physically valid reaction models must be used if it is hoped to use the resulting kinetic constants outside the experimental range of conditions.

In the current investigations of FeS_2 conversion, three models were investigated-- a model assuming overall rate control by solid-state diffusion; a model nth-order in the solid conversion; and a shrinking core model.

EXPERIMENTAL PROGRAM

Single crystals approximately 1 cm in diameter of mineral pyrite from Mexico were ground and screened into four size fractions: 100 x 200, 48 x 100, 28 x 48, and 16 x 28 Tyler mesh (74 x 147, 147 x 295, 295 x 589, and 589 x 991 micrometer). The crushed particles were sharp-edged and angular, similar in appearance to crushed glass. These samples were stored under nitrogen in a dry-box until the time of their use.

Well-formed, cubic crystals up to 5 millimeters on a side were extracted from a pyrophyllite sample from Moore County, North Carolina which contained a large number of embedded crystals.

Pyrite samples were also obtained from Western Kentucky No. 9 coal sample by a two-stage sink/float procedure. The bulk of the organic material was separated by a sink/float in ethylene dibromide (specific gravity = 1.9). The concentrated mineral fraction thus obtained was subjected to a sink/float in bromoform (specific gravity = 2.9). The most dense fraction obtained was essentially FeS_2. Examination by energy dispersive X-ray showed few particles containing elements other than iron or sulfur. (Carbon or lighter elements are not detected by this analysis. Later experiments indicated the presence of 5 to 10 percent carbonaceous impurities.)

Experimental Rate Measurements

The rates of FeS_2 decomposition were measured in a DuPont 951 thermogravimetric analyzer (TGA). The TGA uses a sensitive balance to continuously weigh a sample held in a platinum pan. The sample is enclosed by a quartz tube which is in turn sur- rounded by an electrically heated furnace. A thermocouple located beside the sample pan continuously monitors the temperature of the sample environment. Linear heating rates from 0.1°C to 100°C/min are obtain- able up to a maximum temperature of 1200°C. The quartz tube surrounding the sample permits a continuous flow of reactive gases to be passed over the sample and into a vent system. Sample weight is digitally recorded on magnetic discs at maximum sensitivity of 1 microgram, but background vibration limits the practical sensitivity to about 5 microgram. In addition, under heating conditions, the apparent sample weight can exhibit some drift even if no actual weight loss is occurring (less than 0.3 microgram/°C).

The balance has an ultimate capacity of about 300 milligram, but smaller samples exhibit less internal variation in temper- ature and concentration gradients. For this study, sample weights of about 20 to 30 milligram provided easily measured weight changes while maintaining a thin layer of sample particles in the pan.

Nonisothermal experiments at lower heating rates have the advantage of better resolving reaction events, and of mini- mizing temperature gradients within the sample. The majority of experiments were run at either 2,5, or 8°C/min, with a few runs being made at 1 and 16°C/min.

Most of the experiments were performed in one of three gaseous atmospheres: nitrogen, hydrogen, or carbon monoxide. Airco Grade 2.3 carbon monoxide was used (minimum purity of 99.3 percent; maximum of 150 ppm nitrogen, 600 ppm oxygen, and 50 ppm CO_2.). Airco Grade 5 hydrogen was used. (Minimum purity of 99.99 percent; maximum of 0.5 ppm oxygen; and 2 ppm of water.) The nitrogen used was passed through an Alltech Oxy-trap to reduce residual oxygen content to less than 0.1 ppm. To investigate the effect of gas diffusivity, runs were also made in helium and argon and then compared with the nitro- gen runs. These two gases were also passed through the Oxy-trap prior to use.

Gas flows into the reaction tube were controlled by rotameters within ±5 percent during an individual run. With different gases, flow rates between 50 and 100 cm^3/ min were used. It should be noted that these gas flows are much more than the stoichiometric quantity required to decom- pose the solid sample. Even at the maximum decomposition rate measured from any run, the gas flows were 100 to 200 times the amount required to decompose the solid.

Gas flow through the quartz tube is well within the laminar flow regime--Reynolds numbers of 3 to 10 are typical.

SOLID-STATE DIFFUSION AS A CONTROLLING MECHANISM

In a review article on the chemistry of coal sulfur reactions, Attar (15) argues that the observed rates of FeS_2 decomposition are best explained by a rate-limiting step of solid-state diffusion. The availability of S^{--} ions at the solid surface was believed to control the reaction rate. In gases which react with S^{--}, the concentration of sulfur atoms at the solid surface would be kept low, the driving force for ion diffusion through the solid interior would be increased, and the over-all rate of reaction would go up.

Wüstite (a nonstoichiometric iron oxide) decomposition in hydrogen or carbon monoxide has been shown to be controlled by solid-state diffusion under certain conditions (16, 17). Considering potential similarities between the Fe-O and Fe-S systems, the possibility of a solid-state diffusion controlled reaction was studied.

Single well-formed, cubic crystals of FeS_2 were removed from the pyrophyllite sample, placed in the TGA, and heated at 5°C/min until a portion of the sample had reacted. The sample was then withdrawn from the furnace and cooled to suddenly stop the reaction. Four runs were made--two in hydrogen (corresponding to 19 and 40 percent removal of the first mole of sulfur atoms) and two in nitrogen (corresponding to 14 and 33 percent removal). Each particle was then sliced in two with a diamond saw and examined.

The partially reacted particles con-sisted of a darker layer of product sur-rounding a hard inner core (see Figure 2). The product layer crumbled easily, leaving a dense core which showed the hardness and lustre of the original FeS_2. The crystals were measured before and after reaction and were found to swell considerably--1.76 m^3 of product layer formed per m^3 of core reactant consumed. This overall volume increase contrasts with a decrease in the crystalline molar volume upon reaction; FeS_2 has the larger molar volume (24.5 cm^3/mol versus 18 cm^3/mol for FeS). The net swelling observed is due to an increase in the solid porosity upon loss of the first mole of sulfur.

The solid structure changes abruptly between product and core as strikingly shown in microphotographs of the product/core interface (Figures 3 and 4). A porous product layer of long, rod-like crystals abuts against a nonporous core. This structure is the same for decomposition in hydrogen or nitrogen.

Qualitative composition within the partially reacted particles were measured by energy dispersive X-ray analysis. A small area (approximately 0.1 mm square) was analyzed for the ratio of iron to sulfur signal. In the core, the signal ratio averaged 2.8 S:Fe, and in the product layer 1.8 S:Fe. These measurements are not quantitative, instead, their interest lies in the fact that there was no concentration gradient inside the core. The S:Fe ratio measured just inside the reaction interface was the same as that in the center of the core.

These measurements argue against a reaction model controlled by solid-state diffusion, since if diffusion through the solid were controlling, an increasing gradient in the S:Fe ratio would be ex-pected toward the center of the particle, with a gradual transition from reactant to product.

AN NTH-ORDER REACTION MODEL

Rate Expression

A reaction model often used for homo-geneous reactions is one which assumes the rate is proportional to a temperature dependent rate constant times the concen-tration of each reactant raised to some power:

$$\frac{dN_A}{dt} = - k C_A^m C_B^n. \tag{1}$$

Exponents of 1, 2, or 3 result from theoretical considerations in elementary gas phase reactions (18), but in practice noninteger values of n are the rule, and Equation (1) is of essentially empirical nature.

The concept of an nth-order rate dependence can be extended to solid-gas reactions if some means of specifying solid "concentration" is chosen. Typically, the fraction remaining of the original solid reactant is used as the "concentration" term.

For the decomposition of FeS_2, an irreversible nth-order reaction model yields the following expression for the rate of fractional weight loss.

$$A_{(gas)} + bB_{(solid)} \rightarrow cC_{(gas)} + dD_{(solid)}$$

$$\frac{dW^*}{dt} = - \frac{f\phi}{MW_B} kC_A^m \left[1 - \frac{MW_B}{\phi}\left(1 - \frac{W^*}{f}\right) \right]^n \qquad (2)$$

In this expression, f is the fraction of original sample particles which still remain at any time. It was found necessary to include f because during the FeS_2 decomposition, some particles would fracture and eject themselves from the sample pan causing an abrupt weight loss. The fraction, f, is obtained from the ratio of weight prior to and following a sudden loss.

The factor ϕ represents the grams of total weight lost per mole of solid B consumed. Its value calculated from the reaction stoichiometry is 32.06.

Since all experimental runs were made in pure gases at one atmosphere pressure, the gaseous concentration dependence could not be determined. It was incorporated into an apparent rate constant for each reaction,

$$C_A^m k = k_{ap} = A \exp\left(\frac{-E}{RT}\right) .$$

The value of n in Equation (2) takes on constant values in certain limiting cases--zero for rate control by gas film diffusion and 2/3 for rate control by chemical reaction at a nonporous surface (19). However, for more complicated reactions, the order with respect to solids conversion varies in a complex manner as the reaction proceeds and a single value of n cannot be expected to hold throughout the reaction.

Model Parameters for Hydrogen Atmosphere

Data fit to a linearized version of Equation (2) produce equally good fits for $1 \leq n \leq 2$. To preserve mathematical simplicity, n = 1 was chosen. Larger particle sizes had smaller pre-exponential factors but the same activation energy. As shown in Table 1, the pre-exponential factor, A, was found to vary inversely with respect to particle radius. (Theory would predict linear inverse dependence of A if the rate depended on the external surface area of the particle.) This suggested normalizing all data by defining

$$A^* = \frac{\phi}{MW_B} AR_0$$

where R_0 = 1/2 the geometric average of the screen openings of the particle size cut.

The resulting equation rewritten in terms of fractional conversion of FeS_2 to FeS is:

$$\frac{dX}{dt} = \frac{A^*}{R_0} \exp\left(\frac{-E}{RT}\right)\left(1 - X\right)$$

where for crystalline pyrite from a mineral source,

A^* = 33 x 10^5 m · min^{-1}

E = 45.7 kcal/mol (191 kJ/mol)

X = fraction of FeS_2 converted to FeS.

Model Parameters for Carbon Monoxide Atmospheres

The decomposition of FeS_2 began at slightly lower temperatures in a carbon monoxide atmosphere. Reduction and normalization of the carbon monoxide data gave E = 39.7 kcal/mol (166 kJ/mol) with

A^* = 6.0 x 10^5 m · min^{-1}.

It is interesting to note that the activation energy for decomposition in CO is very close to that in H_2--39.7 versus 45.7 kcal/mol. A statistical t-test indicated these values are different at a 90 percent confidence level, but not at a 95 percent confidence level. It is not known whether the similar values reflect a common underlying mechanism or simply coincidence.

Measurements in Inert Atmospheres

Several runs were made of FeS$_2$ decomposition in nitrogen, but they could not be correlated by an irreversible nth order reaction since two maxima occurred in the rate curve (see Figure 5). Decomposition began around 450°C, increased in rate until 500-520°C, dropped in rate until at about 540°C the rate began to accelerate once more.

It is speculated that the preliminary slowing of the decomposition is due to equilibrium limitation (by sulfur vapor at the reaction interface) of the reaction

$$2FeS_2 \rightleftarrows S_{2_{(g)}} + 2FeS.$$

At low temperatures, diffusion of S$_2$ vapor into the bulk gas could be rapid enough to keep surface concentrations below equilibrium levels, while at higher temperatures, the exponential increase of the equilibrium constant would override any concentration buildup of S$_2$. It is postulated that only in the middle temperature range do the concentrations and equilibrium constant lead to reaction slowing.

If true, inert gases with greater diffusivities should reduce the rate inhibition by product sulfur vapor. In helium ($D_{AB} \approx 2$ to 3 cm^2/s for He-S$_2$ versus 0.6 to 0.9 cm^2/s for N$_2$-S$_2$) a bimodal rate curve was still obtained, but the second peak occurred at significantly lower temperatures. In argon ($D_{AB} \approx 0.5$ to 0.8 cm^2/s) the results were similar to those in nitrogen. Figure 6 shows the rate curves for 100 x 200 mesh (74 x 147 micrometer) mineral FeS$_2$ heated at 5°C/min in the three inert gases.

SHRINKING CORE MODEL FOR LINEAR HEATING RATES

The observation of a porous layer surrounding a solid core suggests the shrinking core model (SCM) as a method of correlating the reaction rate data for FeS$_2$ decomposition.

The SCM is one of the most widely used models for analyzing solid-gas reaction rates ([19], [20]). It postulates a reaction interface inside each solid particle at which the solid composition abruptly changes from reactant to product. During the course of the reaction, this interface recedes into the particle, leaving a product layer of increasing thickness. The overall rate of solid decomposition may be controlled by (a) diffusion of reactant gas from the bulk gas phase to the outer surface of the particle; (b) the diffusion of reactant gas through the layer of solid product; (c) the rate of chemical reaction at the interior interface; or (d) a combination of all these factors. The rate-controlling effect may change as temperature, particle size, and reactant gas concentration are varied.

Physically, the SCM results when the gaseous reactant combines with the solid reactant much faster than the rate at which it can diffuse through the unreacted solid. It can be shown that the SCM is a limiting case of a more general scheme in which diffusion and combination of the reactants are of comparable speed ([18], [21]).

Usually, the SCM is applied to constant temperature experiments. The adaptation to linear heating rates is straightforward so long as spatial temperature gradients can be neglected, and is outlined below.

SCM for Equimolar Counter Diffusion

Consider a reversible reaction:

$$A_{(gas)} + bB_{(solid)} \rightleftarrows C_{(gas)} + dD_{(solid)},$$

e.g., $H_2 + FeS_2 \rightleftarrows H_2S + FeS$

$$CO + FeS_2 \rightleftarrows COS + FeS.$$

Notice that this case produces one mole of gaseous product for each mole of gaseous reactant consumed. This restriction allows equimolar diffusion to be assumed. By also assuming pseudosteady state, i.e., the rate of reactant gas diffusion equals its rate of consumption at the interface, the decomposition rate can be explicitly found for a first-order reaction. (The error associated with the pseudosteady state assumption is discussed by Wen [21]. It is usually negligible for gas-solid reactions.)

The total molar flow of reactant gas passing through the gas film which surrounds a spherical particle is

$$\frac{dN_A}{dt} = -4\pi R^2 h_D \left(C_{A_o} - C_{A_s} \right). \tag{3}$$

For diffusion through the layer of solid product, an effective diffusivity is defined and

$$\frac{dN_A}{dt} = -4\pi r^2 D_{eff} \frac{\partial C_A}{\partial r}.$$

Since $\frac{dN_A}{dt}$ is constant with position at all times, integration yields

$$\frac{dN_A}{dt} = \frac{4\pi D_{eff}}{\frac{1}{R} - \frac{1}{r_c}} \left(C_{A_s} - C_{A_c} \right). \tag{4}$$

For the chemical reaction rate, a first-order, reversible reaction consumes component A at a rate

$$\frac{dN_A}{dt} = -4\pi r_c^2 k \left(C_{A_c} - C_{A_{eq}} \right).$$

The assumption of spatially uniform temperature implies $C_{A_o} + C_{C_o} = C_{A_c} + C_{C_c}$. An equilibrium constant is defined by $K_c = (C_C/C_A)_{eq}$ and elimination of $C_{A_{eq}}$ by substitution yields

$$\frac{dN_A}{dt} = -4\pi r_c^2 k \left(\frac{1+K_c}{K_c} \right) \left(C_{A_c} - \frac{C_{C_o} + C_{A_o}}{1+K_c} \right). \tag{5}$$

Finally, by eliminating the unmeasurable concentrations C_{A_c} and C_{A_s}, the relationship for the consumption of gaseous component A is found:

$$\frac{dN_A}{dt} = \frac{-\left(C_{A_o} - \frac{C_{A_o} + C_{C_o}}{1+K_c} \right)}{\left(\frac{1}{4\pi r_c^2 \left(\frac{1+K_c}{K_c} \right) k} + \frac{1/r_c - 1/R}{4\pi D_{eff}} + \frac{1}{4\pi R^2 h_D} \right)} \tag{6}$$

All that remains is to eliminate the unknown radii, r_c and R, and express the resulting expression in terms of overall weight. The consumption of solid reactant, B, is obtained from the stoichiometry and ρ_B, the bulk molar density of B. The rate at which the unreacted core shrinks is then given by

$$\frac{dr_c}{dt} = \frac{1}{4\pi r_c^2 \rho_B} \frac{dN_B}{dt} = \frac{b}{4\pi r_c^2 \rho_B} \frac{dN_A}{dt}.$$

Provision can be made for the case in which the overall particle diameter varies because the solid product and reactant exhibit different bulk densities. Defining Z as the ratio of product volume to reactant volume, the relationship for overall particle radius is:

$$R^3 = Z R_o^3 + (1 - Z) r_c^3 \tag{7}$$

When all the necessary substitutions are made, the final expression for the rate of fractional weight loss is given by:

$$\frac{dW^*}{dt} = \frac{\frac{-3b\phi f P}{\rho_B MW_B R_o^3 RT} \left\{ Y_{A_o} - \frac{1}{1+K_c} \right\}}{\left\{ \frac{K_c}{r_c^2 k(1+K_c)} + \frac{R-r_c}{R r_c D_{eff}} + \frac{1}{R D_{AC}} \right\}} \tag{8}$$

where $r_c = R_o \left\{ 1 - \frac{MW_B}{\phi} \left(1 - \frac{W^*}{f} \right) \right\}^{1/3}$,

$h_D = D_{AC}/R$ (for a stagnant surrounding gas).

R is given by Equation (7) and other terms are as defined above.

Equation (8) predicts the overall rate at which gaseous A reacts when given experimental concentrations and the parameters governing the rate of the individual steps. Especially notice that the derivation does not restrict Equation (8) to an isothermal experiment; it only requires that there be no spatial variations in temperature at a given time. By specifying a heating profile, the progress of a reaction under a varying temperature program can be found.

After specifying properties, the SCM still contains three adjustable model parameters: the activation energy, E; the Arrhenius pre-exponential factor, A; and a parameter for the effective diffusivity of the product layer, D_{eff}. The particle size (R_0) and heating rate (α) can be varied between experiments.

In practice, the equilibrium constant is also an adjustable parameter to a certain extent. While calculable in theory, Gibbs' free energies from different sources (22, 23, 24) show sufficient variation that K_c can only be approximately calculated. In this paper, the relations cited by Vaughan and Craig (24) were chosen as the most applicable and used to calculate K_c:

$$H_2S + FeS_2 \rightleftharpoons FeS + H_2S$$

$$\ln K_c = 11.38 - 6215/T$$

$$CO + FeS \rightleftharpoons FeS + COS$$

$$\ln K_c = 5.91 - 4720/T$$

The parameter D_{eff} can be estimated if the pore size and geometry are known for the solid product. The photograph in Figure 4 supports a model of uniform diameter, round pores running perpendicular to the surface. For this simplified model, Szekely, et al. (20) summarize relationships for estimating D_{eff}. For a pore diameter of 3 micrometers estimated from Figure 4, and a porosity of 0.6 (calculated from overall volume increase and true crystal densities), the overall value of D_{eff} is determined by Knudsen diffusion. For the gases and temperatures encountered in FeS_2 decomposition, the magnitude of D_{eff} is estimated to be 10^{-2} to 10^{-3} cm^2/s and

only a week function of temperature ($D_{eff} = CT^{1/2}$).

The experimental runs were analyzed using a form of Equation (8) linearized in $\ln A$ and E/R. Upon assuming D_{eff}, the values of A and E/R were calculated by linear least squares. For 100 x 200 mesh (74 x 147 micrometer) particles in hydrogen, the effect of varying D_{eff} is shown in Figure 7. For 100 x 200 mesh particles, physically meaningful values of D_{eff} have only a small effect on the overall fit of the data. This results because small particles do not develop thick enough layers of products to strongly affect the overall rate. The best relationship for D_{eff} was chosen as $D_{eff} = 2 \times 10^{-6} T^{1/2}$ $[m^2/min]$. Decomposition in carbon monoxide atmospheres was analyzed in a similar manner. The rate expressions and resulting parameters for the CO and H_2 runs are summarized in Table 2.

COMPARISON OF MINERAL AND COAL-DERIVED PYRITE

Measurement of Decomposition Rate

The main body of decomposition measurements were performed on FeS_2 obtained from large single crystals of mineral pyrite. It was recognized that the FeS_2 that occurs in coal may exhibit different behavior. A pyrite rich fraction separated from a Western Kentucky No. 9 coal was screened to 100 x 200 Tyler mesh (74 x 147 micrometer) and was decomposed in N_2, CO, and H_2.

In hydrogen and carbon monoxide, the coal-derived pyrite exhibited greater reactivity than the mineral variety. Figure 8 shows the results for comparable samples of coal-derived and mineral pyrite heated in hydrogen at 5°C/min. The coal-derived sample shows some initial weight loss below 400°C because coal impurities in the sample devolatilize to char and gaseous products. Devolatilization weight loss is superimposed as a shoulder on the main rate curve. Otherwise, comparable samples show the same decomposition behavior, only beginning at lower temperatures for the coal-derived material.

In a nitrogen atmosphere, no enhancement of decomposition rate was seen for the coal-derived FeS_2; the decomposition behavior of comparable samples was essentially identical. This supports the postulate of an equilibrium-controlled rate step for decomposition in inert atmospheres. That is, the increased activity of coal-derived FeS_2, observed in other gases, does not increase the overall rate in nitrogen because equilibrium constrains the degree of decomposition, not the chemical reaction rate.

Because both mineral and coal-derived FeS_2 displayed the same activation energy, higher reaction rates for coal-derived material could be correlated by using a reactivity correction factor outside the exponential term of the Arrhenius rate expression. A reactivity multiplier of about 10 was computed for the Western Kentucky coal-derived pyrite.

Microscopic Examination

Figures 9 through 11 show electron microscope photographs of pyrite particles separated from the Western Kentucky No. 9 coal. They display much more varied forms than the angular, "crushed glass" appearance observed for the mineral pyrite samples.

Figure 9 was exceptionally interesting. When seen under a stereoscopic view, the numerous ovals were tall spires extending from the flat surface. (It is suggested that these may have formed by infilling of plant pores during coalification.) Figure 10 shows an otherwise solid crystal which exhibits a porous structure in the lower portion. Figure 11 was judged typical of the majority of particles.

It is interesting to calculate the porosity parameters needed to explain the increased activity of coal-derived pyrite solely in terms of higher surface areas. Particles 100 x 200 mesh are represented as spheres of 105 micrometer in diameter. The corresponding surface area for completely nonporous FeS_2 particles is of the order 0.01 m^2/gm. The surface area thus needed to explain the observed reactivity corrections is of order 0.1 m^2/gm, not at all an unreasonable value. In the past, the high reactivity of coal-derived pyrite compared to mineral samples has been attributed to a different crystalline or chemical form. The current work suggests increased surface area may be at least as important and perhaps the dominant factor in the higher reactivities.

UNANSWERED ISSUES

This paper has not addressed several topics of interest in FeS_2 reactions, and has generated additional areas of interest.

Among them are the detailed behavior of the second stage of desulfurization, $FeS \rightarrow Fe$. Only in hydrogen did this reaction occur at temperatures below 900°C, and the resulting rate curve indicated a very low activation energy for the controlling process. The loss of the second sulfur atom will be a complex phenomena since the starting material is porous and since FeS is actually a nonstoichiometric compound exhibiting a range of compositions.

It would be interesting to investigate the rate in additional reactive atmospheres, particularly steam and unsaturated hydrocarbons such as ethylene, which would be expected to co-evolve during heating of coal. Also, gas mixtures should be investigated to determine possible interactions and the effect of gas phase concentration on decomposition rates.

Finally, more work is needed to determine why coal-derived pyrites exhibit higher reactivities. Is the chemical structure significantly different or is increased surface area mainly responsible?

ACKNOWLEDGEMENT

Research Triangle Institute gratefully acknowledges financial support for the above studies provided by the U.S. Environmental Protection Agency (Grant R804979) and the North Carolina Energy Institute (NCEI-10). Acknowledgements go to J. P. Masten for electron microscopy and energy dispersive X-ray analyses.

NOTATION

A	=	pre-exponential factor in Arrhenius equation for the rate constant.
A*	=	pre-exponential factor in nth order model normalized for particle size $(m \cdot min^{-1})$.

b = stoichiometric coefficient of solid reactant.

c_A = molar concentration of species A $(mol \cdot m^{-3})$.

D_{AC} = binary gaseous diffusion coefficient for A and C $(m^2 \cdot min^{-1})$.

D_{eff} = effective diffusivity of solid product layer $(m^2 \cdot min^{-1})$.

E = activation energy in Arrhenius equation $(cal \cdot mol^{-1})$.

f = fraction remaining of particles originally present at beginning of run.

h_D = external mass transfer coefficient based on concentration $(m \cdot min^{-1})$ $(h_D = D_{AC}/R$ for a stagnant surrounding gas).

k = rate constant given by Arrhenius equation k = A exp (-E/RT).

k_{ap} = apparent rate constant incorporating concentration dependence of gas phase.

K_C = equilibriun constant for reaction under consideration.

MW_B = molecular weight of solid reactant.

N_A = moles of component A.

r_c = radius of unreacted core (m).

R = overall radius of particle (m).

R_o = original radius of particle (m).

R = universal gas constant.

t = time (min).

T = temperature (K).

W = sample weight (gm).

W_o = initial sample weight (gm).

W^* = normalized sample weight = W/W_o.

X = fractional conversion of first mole of sulfur in $FeS_2 \rightarrow FeS$.

Y = gas phase mole fraction.

Z = m^3 of product layer formed per m^3 of core consumed.

Greek

α = heating rate of sample $(K \cdot min^{-1})$.

γ = reactivity correction for coal-derived pyrite.

ϕ = overall weight loss per mole of solid reactant converted.

ρ_B = bulk molar density of unreacted solids $(mol \cdot m^{-3})$.

Subscripts

A, B, C = components A, B, and C.

ap = apparent rate constant.

max = maximum rate of overall weight loss.

o = initial time in W_o, bulk gas composition in C_{A_o}.

s = external surface of particle.
s = external surface of particle.

c = reacting core surface.

eq = equilibrium quantity.

Superscripts

m = reaction order with respect to gaseous reactant.

n = reaction order with respect to solid conversion.

LITERATURE CITED

1. Deurbrouck, A. W., "Sulfur Reduction Potential of the Coals in the United States, "Bureau of Mines Report of Investigations, RI-7633 (1972).

2. Thiessen, G., "Forms of Sulfur in Coal," Chemistry of Coal Utilization, Vol. I, ed. H. H. Lowry, John Wiley & Sons, NY (1945).

3. King, H. M., "The Morphology, Maceral, Association and Distribution of Iron Disulfide Minerals in the Waynesburg Coal at a Surface Mine," M.S. Thesis, West Virginia University, Morgantown, WV (1978).

4. Neavel, R., "Sulfur in Coal: Its Distribution in the Seam and in Mine Products," unpublished Ph.D. Dissertation, Penn State University (1966).

5. Grady, W. C., AIME, Trans., 262; 268-274, (1977).

6. Wandless, A. M., Fuel, 48; 54-62 (1955).

7. Pennsylvania State University, "Characterization of Mineral Matter in Coal and Liquefaction Residues," EPRI 366-1 Annual Report (December 1975).

8. Pennsylvania State University, "Characterization of Mineral Matter in Coal and Liquefaction Residues," EPRI AF 417 Project 366-1 (June 1977).

9. Thompson, F. C., and N. Tilling, J. of Soc. of Indust. Trans., 43: 9; 37T-46T (February 29, 1924).

10. Ozawa, T., J. Thermal Analysis, 9; 369-373 (1976).

11. Kissinger, H. E., J. Research National Bureau of Standards, 57; 217 (1956).

12. Flynn, J. H., and L. A. Wall, J. Research National Bureau of Standards, 70A: 6; 487-523 (November-December 1966).

13. Ozawa, T., J. Thermal Analysis, 2; 301-304 (1970).

14. Ozawa, T., J. Thermal Analysis, 7; 601-617 (1975).

15. Attar, A., Fuel, 57: 4; 201-212 (April 1978).

16. Landler, P.F.J., and K. L. Komarek, Trans. Met. Soc. ASME, 236; 138-149 (February 1966).

17. Leven, R. L., and J. B. Wagner, Trans. Met. Soc. ASME, 233; 159-168 (1965).

18. Carberry, J. J., "Chemical and Catalytic Reaction Engineering," McGraw-Hill, NY (1966).

19. Levenspiel, O., "The Chemical Reactor Omnibook," OSU Book Stores, Inc., Corvallis, OR (1979).

20. Szekely, J., J. W. Evans, and H. Y. Sohn, "Gas-Solid Reactions," Academic Press, NY (1976).

21. Wen, C. Y., Ind. Eng. Chem., 60: 9; 34-54 (September 1968).

22. Perry, R. H., and C. H. Chilton, Chemical Engineers Handbook, 5th ed., Table 3-202, McGraw-Hill, NY (1973).

23. Weast, R. C., ed., Handbook of Chemistry and Physics, 52nd ed., p. D-61, Chemical Rubber Company, Cleveland, OH (1971).

24. Barton, P. B., and B. J. Skinner, in Geochemistry of Hydrothermal Ore Deposits, Holt, Rinehart, and Winston, NY (1978) as cited in Vaughan, D. J., and J. R., Craig, Mineral Chemistry of Metal Sulfides, Cambridge University Press, London (1978).

DAUGHERTY, D.P.

TABLE 1. EFFECT OF PARTICLE SIZE ON A FOR
FIRST-ORDER REACTION

Reacting Atmosphere	Particle Size (Tyler Mesh)	A in $k = A \exp\left(\frac{-E}{RT}\right)$ (min^{-1})
Carbon Monoxide ($E = 39.7$ kcal/mol)	100 x 200	4.3×10^{10}
	16 x 28	0.58×10^{10}
Hydrogen ($E = 45.7$ kcal/mol)	100 x 200	1.7×10^{10}
	48 x 100	1.0×10^{10}
	28 x 48	0.81×10^{10}
	16 x 28	0.57×10^{10}

TABLE 2. SUMMARY OF RATE EXPRESSIONS FOR
$$A_{(g)} + FeS_2 \rightarrow C_{(g)} + FeS$$

Nth Order Model, X = fraction of $FeS_2 \rightarrow FeS$

$$\frac{dx}{dt} = \gamma \frac{A^*}{R_o} \exp\left(\frac{-E}{RT}\right)(1 - X)$$

Mineral FeS_2 in H_2, $\gamma = 1$

$A^* = 33 \times 10^5$ (m $\cdot$ min^{-1})

$E = 45.7$ (kcal $\cdot$ mol^{-1})

Mineral FeS_2 in CO, $\gamma = 1$

$A^* = 6.0 \times 10^5$ (m $\cdot$ min^{-1})

$E = 39.7$ (kcal $\cdot$ mol^{-1})

Shrinking Core Model

$$\frac{dN_B}{dt} = \frac{-\frac{P}{RT}\left(y_{A_o} - \frac{y_{A_o} + y_{C_o}}{1 + K_c}\right)}{\left[\frac{1}{4\pi r_c^2\left(\frac{1 + K_c}{K_c}\right)\gamma A \exp\left(\frac{-E}{RT}\right)} + \frac{1/r_c - 1/R}{4\pi D_{eff}} + \frac{1}{4\pi R^2 h_D}\right]}$$

where:

$$r_c = R_o (1 - X)^{1/3}$$

$$R_3 = ZR_o^3 + (1 - Z)r_c^3$$

$$Z = 1.76$$

$$\frac{h_D R}{D_{AC}} = 1.0$$

Mineral FeS_2 in H_2, $\gamma = 1$

$K_c = \exp (11.38 - 6215/T)$

$A = 1.5 \times 10^{10}$ (m $\cdot$ min^{-1})

$E = 46.4$ (kcal $\cdot$ mol^{-1})

$D_{eff} = 2 \times 10^{-6} T/12$ (m$^2 \cdot$ min^{-1})

$D_{AC} = 93 \times 10^{-8} \frac{T^{1.75}}{P}$ (m$^2 \cdot$ min^{-1})

Mineral FeS_2 in CO, $\gamma = 1$

$K_c = \exp (5.91 - 4720/T)$

$A = 2.3 \times 10^{10}$ (m $\cdot$ min^{-1})

$E = 43.6$ (kcal $\cdot$ mol^{-1})

$D_{eff} = 2 \times 10^{-6} T^{1/2}$ (m$^2 \cdot$ min^{-1})

$D_{AC} = 3.74 \times 10^{-8} \frac{T^{1.75}}{P}$ (m$^2 \cdot$ min^{-1})

Coal-Derived Pyrite from Western Kentucky No. 9 Coal

Above relationships apply with $\gamma \approx 10$.

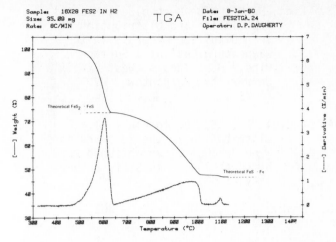

Figure 1. Example output from thermogravimetric analyzer: linear heating of a 16 × 28 Tyler mesh (589 × 991 micrometer) sample of mineral pyrite at 8° C/min in hydrogen at 1 atm pressure. Two major stages of weight loss correspond to $FeS_2 + H_2 \rightarrow$ $FeS + H_2S$ and $FeS + H_2 \rightarrow Fe + H_2S$.

Figure 2. Electron microphotograph of a single crystal of mineral pyrite, partially reacted in hydrogen and sliced in half. Nonporous core is FeS_2 while product layer is FeS. Similar results were seen for partial decomposition in nitrogen.

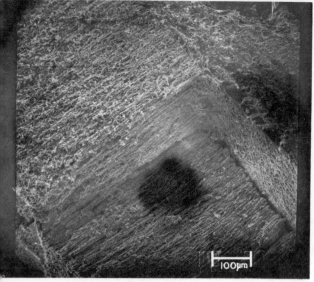

Figure 3. Higher magnification of porous product layer surrounding unreacted core.

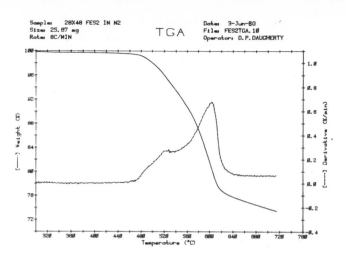

Figure 5. Thermal analysis curve for 28 × 48 mesh (295 × 589 micrometer) mineral pyrite decomposed in nitrogen at 2° C/min. A bimodal rate curve is obtained reflecting, it is believed, an equilibrium constraint on the reaction rate at intermediate conversions.

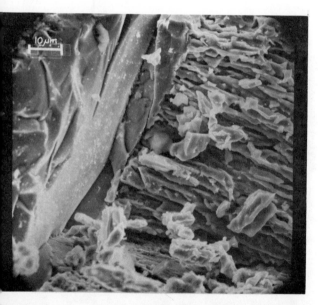

Figure 4. Very high magnification of the reaction interface of nonporous core and porous product layer. Product crystals grow linearly inward as the core recedes.

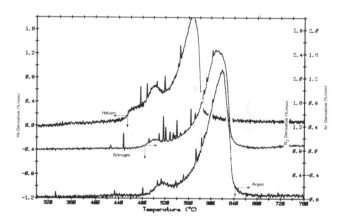

Figure 6. Rate of weight loss of FeS_2 during heating at 5° C/min in three inert atmospheres: N_2, He, and Ar. THe temperature of maximum reaction rate increases as the diffusivity of the gases decrease ($D_{He-S_2} \cong 2\text{-}3$ cm^2/s; $D_{Ar-S_2} \cong 0.5\text{-}9.8$ cm^2/s; and $D_{N_2-S_2} \cong 0.6\text{—}0.9$ cm^2/s). A bimodal rate curve occurs for all three inert atmospheres.

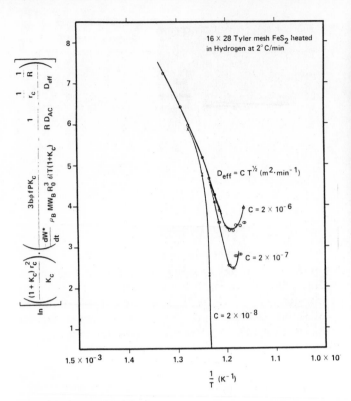

The y-axis label of the plot reads:

$$\ln\left[\left(\frac{(1+K_c)\,r_c^2}{K_c}\right)\left(\frac{dW^*}{dt}\right)\frac{3b_0fPK_c}{\rho_B\,MW_B\,R_0^3\,\delta iT(1+K_c)}\right]\left[\frac{1}{R\,D_{AC}}-\frac{1}{r_c}-\frac{1}{R}\right]\frac{1}{D_{eff}}$$

16 × 28 Tyler mesh FeS$_2$ heated in Hydrogen at 2°C/min

$D_{eff} = C\,T^{1/2}\ (m^2 \cdot min^{-1})$

$C = 2 \times 10^{-6}$

$C = 2 \times 10^{-7}$

$C = 2 \times 10^{-8}$

x-axis: 1.5×10^{-3} 1.4   1.3   1.2   1.1 1.0×10

$\frac{1}{T}\ (K^{-1})$

Figure 7. Linearized plot of shrinking core model for 16 × 28 mesh (589 × 991 micrometer) pyrite in H$_2$. Effect of varying effective product diffusivity for the same data set is shown. For perfect fit of model, curves would be linear over entire range.

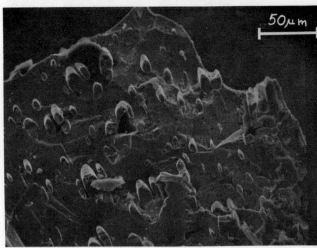

Figure 9. Electron microphotogram of pyrite particle separated from western Kentucky #9 coal by sink/flat in bromoform (specific gravity = 2.9). Upon stereoscopic examination, ovals are tall spires extending outward from the relatively flat surface. Notice recesses and pores in right-hand side of photograph.

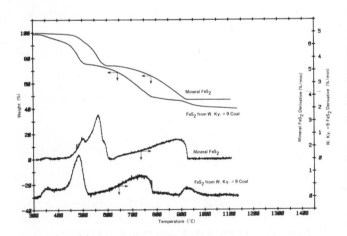

Figure 8. Comparison of mineral and coal-derived pyrite behavior when heated in hydrogen at 5°C/min. Reaction of coal-derived material occurs at lower temperatures. At temperatures below 450°C, weight loss of coal-derived material is partly due to devolatilization of residual coal impurities. Weight loss beginning at about 900°C for coal-derived pyrite may be due to hydrogasification of char impurities.

Figure 10. Electron microphotograph of pyrite from western Kentucky #9 coal. Notice porous nature of otherwise uniform crystal.

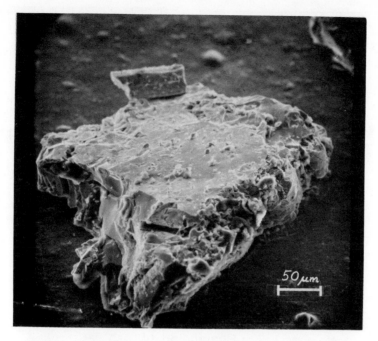

Figure 11. Electron microphotograph of pyrite from western Kentucky #9 coal. Many of the particles examined had the same general characteristics. Smaller particles on top of main particle is also pyrite.

Removal of Organic Sulfur by Low-Temperature Carbonization of Illinois Coals

A linear relationship for total sulfur vs. total iron in low-temperature chars and a slope of the plot corresponding to FeS_2 suggest that pyritic sulfur is fixed by charring while organic sulfur is removed with the volatiles

C.W. Kruse and N.F. Shimp
Illinois State Geological Survey
Champaign, Ill.

One goal of current coal research at the Illinois State Geological Survey (ISGS) is the characterization of coal and coal residues as it relates to the partitioning of major, minor, and trace elements into char and volatile matter during carbonization. A bench-scale, continuous-feed (CF) oven, capable of pyrolyzing coal at a few lb/hr over a range of temperatures, retention times, heating rates, and bed depths, has been designed and built. It is of practical importance that a large proportion of the sulfur atoms remaining in the chars of these Illinois basin coals is derived from pyrite. The char's sulfur appears to be concentrated in clusters close to the regions initially occupied by pyrite, and is more accessible to chemical attack than the diffuse organic sulfur in the coal.

Price and Shieh (1) studied the distribution and isotopic composition of sulfur in coals from the Illinois Basin, and reported that less than 50% of the organic sulfur in high-sulfur coals is of plant origin. They suggested that the plants that formed the low- and high-sulfur coals grew in similar terrestrial environments and that the introduction of a marine-source sulfur responsible for the high-sulfur coals was a secondary event. Smith and Batts (2) reported similar results in studies on a seam of Australian coal that was separated from the overlying marine deposits by some 100 ft (30 m) of sandstone, siltstone, and shale. They concluded that

a considerable period of time elapsed between the end of the coal-forming stage and invasion by the sea. If the secondary event that introduced marine-source hydrogen sulfide into Illinois coals occurred late in the coalification process, the organic sulfur might be loosely attached to pore walls.

Attar (3) describes a method for determining the distribution of organic sulfur functional groups in coal, based on the evolution of hydrogen sulfide as a function of temperature during hydrogenation. This method attempts to quantitatively determine thiolic, thiophenolic, aliphatic-sulfidic, aryl-sulfidic, and thiopheneic organic sulfur. More than 80% of the organic sulfur was accounted for in three coals analyzed, but the method accounted for only 11% of the organic sulfur in an Illinois No. 6 Coal and 46.5% in a Kentucky No. 9/14 blend (Kentucky No. 9 corresponds to Illinois No. 5). If organic sulfur compounds in the Illinois No. 5 and No. 6 Coals were readily lost with volatile components, they might escape without detection. Adsorbed hydrogen sulfide might not give peaks.

Longanbach and Bauer (4) report analyses of flash pyrolysis chars before and after partial combustion. A char produced by charring a subbituminous coal at 870°C lost more than 50% of its sulfur when about 10 to 12% of its weight was lost in a partial combustion at 1040°C. A char produced from a high-volatile bituminous coal at 720°C lost

80% of its sulfur when it was partially combusted at 870°C. The weight loss was 12%. These results would be expected if a part of the char's sulfur were concentrated in easily accessible surface layers removed by oxidation.

Attar (5) concluded, from material balances for the charring of coal under unspecified conditions, that some of the pyrite must have been reduced to elemental iron even at temperatures below 700°C. Calculations of a material balance by Tipton (6) for acid leaching of a flash pyrolysis char from an Occidental Research Corporation (ORC) process showed more gram-atoms of iron and calcium were removed than gram-atoms of sulfur. Analyses reported on the feed coal and char show that all the iron in the feed coal is accounted for in terms of sulfide, pyrite, and iron sulfate, but only two-thirds of the iron is accounted for by these forms in the flash pyrolysis char. The excess iron and calcium extracted over that shown to be in sulfides was assumed to be in the form of oxides. Woolhouse (7) reported a similar imbalance in iron-sulfur stoichiometry for chars produced in the absence of air at 600°C from 5-g quantities of coal. He concluded that the ferrous sulfide formed from pyrite was undergoing a secondary reaction with the coal.

The literature of coal devolatilization through 1975 has been reviewed by Anthony and Howard (8). Literature about the reactions of sulfur in coal-gas reactions has been reviewed more recently by Attar (5). The incorporation of inorganic sulfur from pyrite into the organic matrix by heating of coal with pyrite has been demonstrated by researchers (9, 10, 11). Attar, Corcoran, and Gibson (12) suggest that gaseous hydrogen sulfide resulting from the decomposition of the pyrite reacts with unsaturated organic molecules.

MATERIALS AND METHODS

A schematic diagram of the charring oven is shown in Figure 1. The conveyor is made of close-fitting, overlapping, stainless-steel trays, which form a flatbottomed trough 12.7 cm wide and 2.5 cm deep. Heat is provided by sheathed, electrical resistance elements located above and below the conveyor. Volatilized products are fed through an electrically heated, flue-gas superheater into a gas-fired combustion zone to avoid disposal problems that occur when tar and gaseous products are not collected.

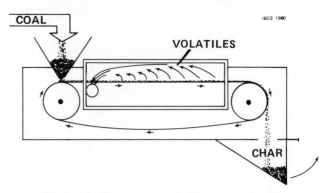

Figure 1. Continuous-feed charring oven.

A weighed quantity of coal is charged between two metal markers that assist the detection of these points as they reach the product bin. Coal is fed before and after the measured charge to maintain a uniform gaseous atmosphere and constant temperature profile throughout the test. The depth of the bed, the residence time, and the properties of the coal determine the temperature differential from inlet to exit of the oven. One hundred fifty tests have been made in this oven at exit temperatures up to 750°C with few mechanical problems.

Chars were leached at reflux temperature for 30 min with 5 N hydrochloric acid, and were recovered by filtration. The filter cake was rinsed with dilute hydrochloric acid, with water, and then with a little acetone. Acid-leached char was hydrotreated in a quartz vessel fitted with a stirrer through a quartz tube that extended to near the hydrogenation zone. The gas inlet through the stirrer assembly directed the incoming hydrogen to the stirred zone. The hydrogen rate was 30 mL/min for a 15 g charge.

Proximate and ultimate analyses of coal and coke were by American Society for Testing and Materials (ASTM) methods D-3172 and D-3176, except for carbon and hydrogen analysis, which was by International Organization for Standardization test 609. ASTM method D-3175 for volatile matter was run on 50:50 mixtures of a coal of known volatile content and a sparking char, when tests on the char alone sparked beyond control. Chlorine determinations were done by ASTM method D-2361, and calorific values were determined by ASTM method D-2015. Iron was determined on ash by atomic absorption using the procedure described by Ruch et al. (13).

Sulfur determinations were made on some

chars by a modification of the oxygen flask method ([14]) that took less time but was less accurate. Finely ground char samples were wetted with isopropyl alcohol to assist burning. The peroxide solution containing the combustion products was diluted to exactly 100 mL and, after the temperature was stable, the pH was determined to three decimal places. Sulfur was calculated from the hydrogen ion concentration, assuming complete ionization of the sulfuric acid.

The forms of sulfur in coal were determined by ASTM method D-2492. A sulfide form of sulfur was added to D-2492 when chars were analyzed. Sulfide determinations were a variation of the evolution method for determining sulfur in iron, steel, ores, cinders, sulfides, and metallurgical products ([15]).

The internal surface areas (ISA) were determined in an apparatus described by Thomas and Damberger ([16]), and the calculations were made using the BET equation ([17]). Adsorption data using nitrogen and carbon dioxide at -196°C and -77°C, respectively, were obtained by a dynamic sorption method. The CO_2 ISA represents the total internal surface area, whereas the N_2 ISA represents only the internal surface areas of pores with diameters greater than 4 to 5 Å.

RESULTS AND DISCUSSION

The sulfur in most Illinois coals exists predominantly in three forms: (a) pyritic sulfur; (b) varying amounts of sulfate ion, depending on the degree of pyrite oxidation; and (c) organic sulfur. The standard test method for forms of sulfur in coal, ASTM method D-2492, does not identify the organic sulfur compounds. The fraction of the total sulfur that exceeds that which can be identified as sulfatic sulfur and pyritic sulfur is designated as organic sulfur ([18]). Any sulfides that release hydrogen sulfide by reacting with hydrochloric acid are reported as organic sulfur according to ASTM method D-2492. If the sulfidic sulfur as well as the sulfatic and pyritic sulfur are determined, then a modified D-2492 can be applied to coke or char, but the organic sulfur remains an indirect determination. The organic compound in the char may not be in the same compounds as those in the parent coal and they need not have the same distribution. Scanning electron microprobe studies show that the organic sulfur of most coals is distributed throughout the macerals, and that the inorganic sulfur appears in clusters

of iron sulfide compounds such as pyrite and ferrous sulfide ([19]). This distributional difference is the basis for electron microprobe methods for direct determinations of organic sulfur ([20], [21], [22], [23]) in coal.

Atoms need to approach one another for chemical reactions to occur, and the fraction of the atoms of a solid exposed to chemical attack and the rate at which they can be attacked are related to surface area and porosity. Therefore, the distribution pattern of atom types in the solid cannot be ignored. A material with low surface area and having a minor element concentrated on its surfaces may expose a higher proportion of these minor atoms to chemical attack than a material with a high surface area of the same composition and having the minor element distributed uniformly throughout the solid.

Figure 2 shows areas as a function of pyrolysis temperatures for chars of three high volatile bituminous coals. These preliminary batch results from 5-g quantities of coal influenced decisions regarding the design of a continuous-feed charring oven, in that they suggested temperatures below 700°C would produce the highest surface areas per unit of coal charged.

Continuous-Feed, Thin-Bed, Counter-Flow Charring

The configuration for removal of volatile matter from the charring oven was dictated by the desire to minimize the interaction of volatiles that are generated at a given temperature with char that has reached

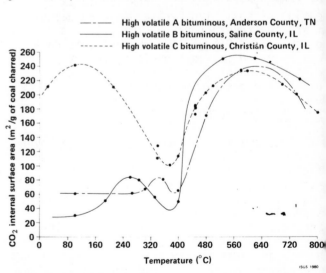

Figure 2. Surface area as a function of charring temperature.

higher temperature. It was also desirable to minimize the deposition of products that may form when volatiles produced in one stage of volatilization react with volatiles produced in another stage. To attain these objectives, it was helpful to position the removal port for the volatile matter near the entrance to the oven so that the volatiles were moved in a gaseous flow opposite to that of the coal bed. The thickness of the bed also affects the interaction of volatile compounds with one another. Appreciable temperature gradients are developed within thick beds heated externally. Consequently, charring a bed of appreciable depth involves passage of the volatiles, which are produced from interior regions late in the charring sequence, through the surface char, which has already produced a part of its volatiles. These surface layers may then produce reactive volatile components, which quickly crosslink the lighter components. Preliminary batch experiments showed that relatively soft char was produced in thin beds from the same coal that gave relatively hard char in bed depths of 25 mm and more.

Properties of Continuous-Feed (CF) Chars

The analyses of CF chars showed a wide variation in the amount of pyritic sulfur remaining in the char (Table 1). Analyses of the feed coals are given in Table 2. For

tests at 600°C and 650°C, a correlation is seen between bed depth and the percentage of the pyrite that is converted to acid-soluble iron. At these temperatures the internal surface areas, measured by nitrogen adsorption, are highest for thin-bed chars. At 700°C the decomposition of the pyrite is uniformly high, and the nitrogen surface areas of the samples appear to be related to residence time, as well as to bed depth.

The ratio of the total sulfur to the total iron in eleven thin-bed chars produced from coals mined in four counties in Illinois is remarkably constant, as shown in Table 3 and plotted on Figure 3. The slope of the line corresponds to 2.03 sulfur atoms per iron atom. The sulfur-to-iron atomic ratios in the four feed coals ranged from 3.1:1 to 5.4:1, and the organic sulfur varied from 45 to 62% of the total sulfur. The reproducibility of these results for each coal is good. Iron appears to control the amount of sulfur remaining in the char. The sulfur that iron controls should be the sulfur combined with it initially. The results are those to be expected if most of the organic sulfur in the coal is contained in the volatile products released during pyrolysis and if the sulfur from the pyrite is retained in the char.

We have previously cited the literature

Table 1. Continuous-Feed Char

Test no.	Anal. no.	Coal source Illinois co.	Size fed. mesh	Oven exit (°C)	Residence time (min)	Bed depth (mm)	Yield, weight (%)	Surface area			Sulfur			Pyrite converted to acid sol. iron (%)
								CO_2 (m^2/g)	N_2 (m^2/g)	Co_2/N_2 ratio	Total (%)	Pyritic (%)	Sulfidic (%)	
F-122	C20875	Fulton	-12 x 40	600	15	5	62.2	285	21.5	12.3	3.07	1.00	.09	63
F-123	C20876	Fulton	-12 x 40	600	15	12	64	279	8.9	31.3	3.82	2.29	.13	12
F-134	C20871	Saline	-12 x 0	650	15	3	62.2	306	21.0	14.6	1.83	0.13	.12	90
F-135	C20872	Saline	-12 x 0	650	15	8	63.7	303	15.2	19.9	1.95	0.39	.19	68
F-136	C20873	Gallatin	-12 x 0	650	15	2	64.0	285	12.0	23.7	3.20	0.42	.52	85
F-137	C20874	Gallatin	-12 x 0	650	15	9	66.6	277	2.7	102.6	3.92	1.64	.56	39
F-138	C20806	Gallatin	-60 x 0	650	15	3	65.9	279	20.3	13.7	3.01	0.28	.51	90
F-140	C20807	Gallatin	-60 x 0	650	18	18	69.6	232	3.3	70.3	3.31	1.68	.26	34
F-141	C20906	Gallatin	-60 x 0	700	15	4	66.0	292	6.1	48	2.90	.10	.63	96
F-144	C20909	Gallatin	-60 x 0	700	15	3	65.8	305	6.4	47.6	3.02	.11	.64	96
F-146	C20911	Gallatin	-60 x 0	700	22	13	67.3	307	11.5	26.7	3.06	.08	.71	97
F-142	C20907	Saline	-60 x 0	700	15	4	66.3	322	16.4	19.6	1.78	.06	.20	95
F-147	C20912	Saline	-60 x 0	700	22	13	69.0	335	21.6	15.5	1.94	.06	.24	95

Table 2. Forms of Sulfur and Iron in Feed Coals

Source Illinois county	Seam	Sulfur (S)				Iron (Fe)				Atomic ratio S_T/Fe_T
		Total (%)	Sulfatic (%)	Pyritic (%)	Organic (%)	Total (%)	Sulfate[a] (%)	Pyrite (%)	Other[b] (%)	
Franklin	6	1.06	.203	.30	.56	0.59	.24	.26	.09	3.1
Saline	6	2.46	.092	.85	1.52	.79	.14	.74	-.09	5.4
Fulton	5	3.29	.023	1.67	1.60	1.49[c]	.03	1.46		3.8
Gallatin	5	3.28	.032	1.78	1.47	1.65	.04	1.55	.06	3.5

[a] Sulfate iron calculated from sulfatic sulfur assuming $Fe_2(SO_4)_3$.
[b] Total iron less sulfate and pyrite.
[c] Estimate, sum of sulfate and pyrite iron.

concerning the incorporation of pyritic sulfur into coal; no one appears to have suggested the conversion to be quantitative. It might approach a quantitative reaction if some type of coordination complex were involved—one formed by a concerted transfer of sulfur to an organic molecule to produce the organic ligand. Coordination compounds of iron are reported to form readily by the reaction of diphenylacetylene and sulfur with iron in refluxing toluene (24). The formation of iron salts of substituted thiophenols is also a possibility; a reaction of pyrite with other mineral constituents is still another possibility. Most of the sulfur in the char reports as organic; however, its true identity may be mineral matter or even elemental sulfur. Its only characteri-

zation is that it is not sulfatic, pyritic, or sulfidic sulfur.

The total iron is accounted for in the feed coals as pyrite and small amounts of sulfate (Table 2). Pyrite is generally assumed to produce pyrrhotite as its major decomposition product during charring, but analyses of the thin bed chars (Table 3) show that less than one-half of the iron in the char can be accounted for as sulfate, pyrite, or pyrrhotite. These chars were produced in a reducing atmosphere; it is unlikely that they were metallic oxides unless they were formed at room temperature after the char was removed from the oven. The possibility that sulfide determinations might not be accurate because the chars might

Table 3. Forms of Sulfur and Iron in Thin-Bed Chars

Test no.	Coal source Ill. co.	Sulfur					Iron				Other	
		Total (%)	Sulfatic (%)	Pyritic (%)	Sulfidic (%)	Organic (%)	Total (%)	Sulfate[a] (%)	Pyrite (%)	Sulfide[b] (%)	(%)	(%)[d]
CF-136	Gallatin	3.20	.089	.42	.52	2.18	2.80	.10	.37	.79	1.54	(55)
CF-144	Gallatin	3.02	.043	.11	.64	2.23	2.62	.05	0.10	.98	1.49	(57)
CF-138	Gallatin	3.01	.182	.28	.51	2.04	2.50	.21	0.18	.78	1.33	(53)
CF-141	Gallatin	2.90	.033	.10	.63	2.14	2.43	.04	.09	.96	1.34	(55)
CF-132	Fulton	2.59	.057	.31	.04	2.18	1.90	.07	.27	.06	1.50	(79)
CF-133	Fulton	2.56	.054	.32	.038	2.15	1.90	.06	.28	.06	1.50	(79)
CF-134	Saline	1.85	.039	.13	.12	1.56	1.37	.05	.11	.18	1.03	(75)
CF-142	Saline	1.78	.012	.06	.20	1.51	1.37	.01	.05	.31	1.00	(73)
CF-69	Franklin[c]	.76	.036	.18	.002	.54	.63	.04	.16	.00	.43	(68)
CF-68	Franklin[c]	.73	.036	.21	.02	.46	.62	.04	.18	.03	.37	(60)
CF-70	Franklin[c]	.70	.044	.09	.002	.56	.62	.05	.08	.00	.49	(79)

[a] Sulfate iron calculated from sulfatic sulfur assuming $Fe_2(SO_4)_3$.
[b] Sulfide iron calculated from sulfidic sulfur assuming Fe_7S_8.
[c] Tests in oven before it was enclosed to exclude all air.
[d] Percent of total iron not accounted for as sulfate, pyrite and sulfide.

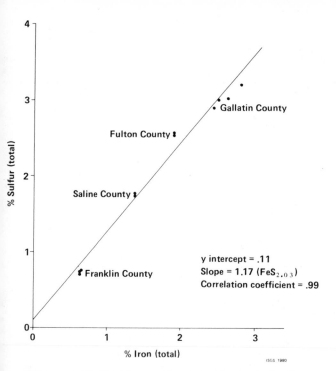

% Sulfur (total)

% Iron (total)

Gallatin County

Fulton County

Saline County

Franklin County

y intercept = .11
Slope = 1.17 (FeS$_{2.03}$)
Correlation coefficient = .99

ISGS 1980

Figure 3. Sulfur-iron ratios for thin-bed chars.

be trapping a part of the hydrogen sulfide released by hydrochloric acid was considered The average of seven sulfide determinations on the CF-141 char was 0.634%, and the standard deviation was 0.023. The char was then saturated with hydrogen sulfide gas at room temperature, and sulfide determinations were repeated. Four analyses averaged 0.635%, and the standard deviation was again 0.023. Three tests showed that the char adsorbed hydrogen sulfide equal to 1.13% of its weight, and the standard deviation was 0.051 This amounts to one-half the sulfur that was lost during charring. None of the adsorbed hydrogen sulfide escaped during the refluxing with 6 N hydrochloric acid. These results, together with Attar's evidence that the organic sulfur in Illinois No. 5 and No. 6 coals may not be one of the expected organic types, raise the possibility that most of the so-called organic sulfur in these coals is adsorbed hydrogen sulfide or low-boiling organic sulfides.

If much broadly distributed organic sulfur is driven off with volatiles during charring and if most of the remaining sulfur is fixed in the vicinity of the initial location of pyrite particles, the distribution of sulfur in char should approach the distribution of the mineral matter. Physical methods of beneficiation that would not work on coal should now be expected to produce a char fraction that is very low

in sulfur. Some success was realized in separating char particles of differing mineral contents with a high tension separator manufactured by Carpco Research and Engineering. A small amount of a low-mineral char having only 0.3% sulfur was separated from char CF-126, which was produced from a Fulton County coal that initially had 1.8% organic sulfur.

The amount of volatile matter produced during the thin-bed charring plus that remaining in the char, as determined by ASTM-3175 at 950°C, was very close to the amount determined initially on the coal in a standard 7-min ASTM test at 950°C (Table 4). Thin-bed charring appears to be producing a fraction of the ASTM volatiles, leaving the remainder available for subsequent removal at higher temperatures.

Desulfurization of Continuous-Feed, Thin-bed Chars

Some desulfurization techniques, which are cited in the literature, are reported to remove some of the organic sulfur from coal. We are currently applying a few of these techniques to the CF chars to determine whether any are more effective on chars than coal. The low-temperature chlorinolysis method described by Jet Propulsion Laboratory (JPL) workers (25) was applied to coal from Macoupin County, Illinois, and to its char (Figure 4). The oxygen flask method was used for these sulfur determinations.

In both tests, 58% of the sulfur in the feed coal was removed. No advantage was apparent in incorporating an intermediate charring step. JPL's last step is dechlorination by heating at 300° to 550°C. When this step was replaced by hydrodesulfurization at 750°C to 800°C and ambient pressure, there was a dramatic increase in the amount of sulfur removed. Results from a series of tests on coal mined in Fulton County, Illinois, are summarized in Figure 5. The tests with chars CF-121 through CF-124 suggest bed depth to be a factor. Chars CF-129 and CF-130 were prepared from a float fraction of the same coal. Tests with these chars gave similar results even though the charring temperatures differed by 150°C. CF-121 char (2.6% sulfur) was also subjected to hydrodesulfurization treatment for one hr without prior chlorination. The product contained 2.8% sulfur. Hydrogen

Table 4. Material Balance of Volatile Matter

Test no.	Bed depth (mm)	Exit temp (°C)	Feed coal				Volatiles produced[b] (%)	Volatiles in char		Ne chan (%
			Illinois county	Iron (%)	Moisture (%)	Volatiles[a] (%)		Calculated[c] (%)	Determined[d] (%)	
CF-136	2	650	Gallatin	1.65	4.4	36.9	31.6	5.3	7.0	+1.
CF-132	2	650	Fulton	1.23	12.3	35.2	24.3	10.9	10.0	-0.
CF-133	2	600	Fulton	1.23	12.3	35.2	21.2	14.0	15.3	+1.
CF-68	2	550	Franklin	.59	12.4	32.5	16.2	16.3	17.6	+1.
CF-70	2	580	Franklin	.59	12.4	32.5	17.5	15.0	15.1	+ .
CF-144	3	700	Gallatin	1.65	2.4	37.1	31.8	5.3	4.9	- .
CF-134	3	650	Saline	.79	4.9	35.7	32.9	2.8	6.6	+3.
CF-69	3	540	Franklin	.59	12.4	32.5	18.2	14.3	14.0	- .
CF-138	3	650	Gallatin	1.65	3.5	37.1	30.6	6.5	5.8	- .
CF-142	4	500	Saline	.79	3.0	35.7	30.7	5.0	5.1	+ .
CF-141	4	700	Gallatin	1.65	2.4	37.1	31.6	5.5	5.3	- .
CF-135	8	650	Saline	.79	4.9	35.7	31.4	4.3	7.3	+3.
CF-146	13	700	Gallatin	1.65	2.4	37.1	30.3	6.8	5.0	-1.
CF-147	13	700	Saline	.79	3.0	35.7	28.0	7.7	5.7	-2.
CF-123	15	600	Fulton	1.48	11.6	35.9	24.4	11.5	10.2	-1.
CF-140	18	650	Gallatin	1.65	2.4	37.1	29.0	8.1	10.7	+2.

[a] ASTM at 950°C.
[b] Percentage weight loss in charring minus percentage moisture.
[c] ASTM volatiles in the feed coal minus the amount produced in charring.
[d] Amount of ASTM volatiles in char as a percentage of the initial feed.
[e] Total volatiles removed in two steps minus the one step ASTM volatiles in coal.

removed nonsulfur components faster than sulfur.

Analyses of chars after the chlorination in a slurry of 1,1,1-trichloroethane and water at 70°C corresponded closely to those of the same char after leaching with an amount of hydrochloric acid equal to that produced by the reaction of chlorine with water. The oxidizing action of the chlorine may be needed to effect dissolution of the pyrite in coal, but little pyrite remains in many of the CF chars. Chlorinolysis was replaced in subsequent tests by acid leaching of the chars with 5 N hydrochloric acid at its boiling point for 30 min. Results for a number of tests are summarized in Table 5. Factors in charring that result in a high removal of the iron during acid leaching appear to control the lower level to which sulfur can be reduced by hydrodesulfurization, as reported by Occidental Research Company workers (6).

A complete analysis of a desulfurized char and its precursors appears in Table 6.

About half the ash is removed by acid leachin so that the ash content of desulfurized char remains approximately the same as that of feed coal. Calorific values remain essentially that of the feed coal and there is no evidence that chlorine content is increased.

Anthony and Howard (8) distinguished the Char Oil Energy Development (COED) process

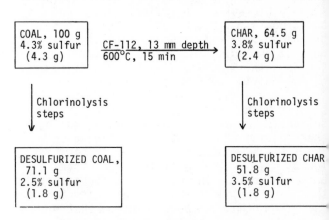

Figure 4. Chlorinolysis of coal and char.

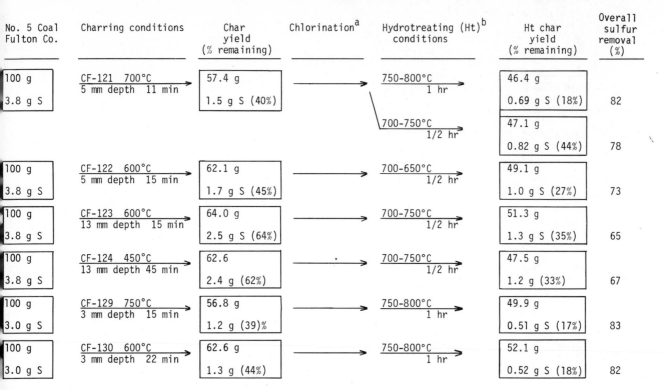

No. 5 Coal Fulton Co.	Charring conditions	Char yield (% remaining)	Chlorination[a]	Hydrotreating (Ht)[b] conditions	Ht char yield (% remaining)	Overall sulfur removal (%)
100 g 3.8 g S	CF-121 700°C 5 mm depth 11 min	57.4 g 1.5 g S (40%)	→	750-800°C 1 hr	46.4 g 0.69 g S (18%)	82
				700-750°C 1/2 hr	47.1 g 0.82 g S (44%)	78
100 g 3.8 g S	CF-122 600°C 5 mm depth 15 min	62.1 g 1.7 g S (45%)	→	700-650°C 1/2 hr	49.1 g 1.0 g S (27%)	73
100 g 3.8 g S	CF-123 600°C 13 mm depth 15 min	64.0 g 2.5 g S (64%)	→	700-750°C 1/2 hr	51.3 g 1.3 g S (35%)	65
100 g 3.8 g S	CF-124 450°C 13 mm depth 45 min	62.6 2.4 g (62%)	→	700-750°C 1/2 hr	47.5 g 1.2 g (33%)	67
100 g 3.0 g S	CF-129 750°C 3 mm depth 15 min	56.8 g 1.2 g (39)%	→	750-800°C 1 hr	49.9 g 0.51 g S (17%)	83
100 g 3.0 g S	CF-130 600°C 3 mm depth 22 min	62.6 g 1.3 g (44%)	→	750-800°C 1 hr	52.1 g 0.52 g S (18%)	82

[a]Chlorination at 70 C in a slurry of 1,1,1-trichloroethane and water.
[b]Hydrogen at 30 mL/min at ambient pressure oven 5 g char.

Figure 5. Desulfurization by charring-chlorinolysis-hydrodesulfurization.

and Occidental Research Corporation's (OCR) Flash Pyrolysis process (formerly the Garrett process) from other potential coal conversion processes in that they: (a) partition the coal into hydrogen-rich fractions and carbon-rich fractions by pyrolysis in a reducing atmosphere; (b) do not add hydrogen from an external source; (c) operate at essentially atmospheric pressure; and (d) depend on utilization of the by-product char for their economic viability. These processes produce a distribution of volatile products and solids in keeping with the hydrogen-carbon balance of coal. The current trend is toward commercialization of coal conversion processes that require economically and thermodynamically costly hydrogen from water to produce gas or liquid products. The trend could ultimately shift to processes that produce products in a proportion of physical states dictated by coal's hydrogen deficiency, if cost-effective processes for the removal of sulfur from coal and its solid residues become a reality. Success in desulfurization of a solid coal depends on the location and forms of sulfur compounds in the coal and in the char as a function of devolatilization.

CONCLUSIONS

Many of the conclusions drawn from the results of this study are circumstantial and need confirmation by other methods. The results do, however, support the following interpretations:

1. A high percentage of the organic sulfur initially in Illinois No. 5 and No. 6 Coals may be removed by charring of thin beds of coal.

2. New forms of sulfur, reported as organic sulfur in the analysis scheme generally used today, are produced by a reaction of pyrite with coal constituents at temperatures below 600°C.

3. Pyrrhotite is not the major product of the decomposition of pyrite during charring of thin beds, despite the high proportion of the iron that is leachable with hydrochloric acid.

4. The sulfur remaining in a char may be concentrated in regions initially occupied in the coal by pyrite and may

Table 5. Desulfurization of Continuous-Feed Chars

Char used	Coal source Ill. co.	Acid leached char[a]			Hydrotreated char[b]			Overall sulfur removal from coal (%)	Charring conditions		
		Yield[c] (%)	Sulfur (%)	Iron (%)	Yield[c] (%)	Sulfur (%)	Iron (%)		Temp (°C)	Depth (mm)	Time (min)
CF-146	Gallatin	65.6	2.14	.30	55.1	.80	.31	87	700	13	22
CF-144	Gallatin	64.7	2.26	.33	53.6	.86	.46	86	700	3	15
CF-141	Gallatin	65.7	2.07	.30	53.8	.87	.43	86	700	4	15
CF-137	Gallatin	64.9	3.27	1.57	51.7	1.65	1.74	74	650	9	15
CF-136	Gallatin	61.1	2.42	.55	49.8	.95	.57	86	650	2	15
CF-131	Fulton	56.7	3.04	.38	46.7	.63	.43	91	700	4	15
CF-130	Fulton	63.9	1.75	.34	48.1	.55	.50	92	600	4	22
CF-129	Fulton	57.4	1.75	.22	46.2	.62	.60	91	750	4	15
CF-147	Saline	69.7	1.47	.19	60.8	.58	.23	86	700	13	22
CF-142	Saline	67.0	1.59	.21	58.4	.69	.47	84	700	4	15
CF-68	Franklin	71.4	.64	.31	52.8	.43	.42	78	550	2	22

[a] 30 min leach with refluxing 5 N hydrochloric acid.
[b] 800°C, 2 hr, 30 ml/min hydrogen, 15 g sample.
[c] Yield from coal.

therefore be more accessible to attack than is the organic sulfur of coal before pyrolysis.

ACKNOWLEDGMENTS

This work was supported in part by funds from the United States Environmental Protection Agency through contract US EPA 68-02-2130. The authors acknowledge contributions by R. J. Helfinstine, mechanical design and coal selection; J. D. Cooper, char preparation, equipment design, and chemical treatments; R. F. Fairfield, coal preparation and charring; L. B. Kohlenberger, J. S. Hendricks K. D. Baer, J. K. Bartz, and T. D. Norden, analysis; L. A. Khan, electrostatic separation; M. M. Chou, coordination complex con-

Table 6. Desulfurization of Saline County, Illinois, No. 6 Coal

Test[a]	Unit	Coal	Char CF-147	Acid-leached char	Desulfurized char
Yield from 100 coal	g	100	69	69.7	60.8
Moisture	%	4.9	.6	3.4	.4
Volatile matter	%	35.7	8.2	9.5	3.8
Fixed carbon	%	51.4	79.3	79.0	86.8
H-T ash	%	8.0	11.9	8.1	9.0
Hydrogen	%	5.36	1.95		1.06[c]
Carbon	%	70.42	80.57		87.42[c]
Nitrogen	%	1.48	1.59		1.32[c]
Oxygen	%	12.34	2.07		.62[c]
Total sulfur	%	2.40	1.94	1.47	.58
Sulfatic sulfur	%	.089	.011	<.03	.001
Pyritic sulfur	%	.83	.06	.03	.02
Sulfidic sulfur	%	b	.24	b	b
Organic sulfur	%	1.48	1.63	1.44	.56
Total iron	%	.79	1.28	.19	.23
Total chlorine	%	.015	.018		.08[c]
Calorific value	kJ/kg (Btu/lb)	29380 (12630.)	29890 (12849.)	28900 (12423.)	30101 (12941.)

[a] Analyses as received basis.
[b] Sulfidic sulfur not determined.
[c] Analyses from another test.

siderations; R. H. Shiley, acid leaching suggestions; and J. Thomas, Jr., surface area measurements.

LITERATURE CITED

1. Price, F. T., and Y. N. Shieh, *Economic Geology*, 74, 1445-1461 (1979).

2. Smith, J. W., and B. D. Batts, *Geochim. et Cosmochim. Acta*, 38, 121-133 (1974).

3. Attar, A., "Sulfur Groups in Coal and their Determinations," in *Analytical Methods for Coal and Coal Products*, vol. 3, p. 612, C. Karr [ed.], Academic Press, Inc., New York (1979).

4. Longanbach, J. R., and F. Bauer, "Fuels and Chemicals by Pyrolysis," in *Industrial and Laboratory Pyrolyses*, p. 476, ACS Symposium Series, No. 32, Am. Chem. Soc., Washington, D.C. (1976).

5. Attar, A., *Fuel*, 57, 201 (1978).

6. Tipton, A. B., "Improved Hydrodesulfurization of Coal Char by Acid Leach," in *Coal Desulfurization*, p. 280, ACS Symposium Series, No. 64, T. Wheelock, [ed.], Am. Chem. Soc., Washington, D.C. (1977).

7. Woolhouse, T. G., *Fuel*, 14, 259-264, 286-295 (1935).

8. Anthony, D. B., and J. B. Howard, *AICHE Journal*, 22, 625-656 (1976).

9. Powell, A. R., *J. Ind. Eng. Chem.*, 12, 1069 (1920).

10. Eaton, S. E., R. W. Hyde, and B. S. Old, *Am. Inst. Mine, Met., Eng., Iron and Steel Div. Metal Tech.*, 19, 343 (1948), Tech. Bull. 2453.

11. Cernic-Simic, S., *Fuel*, 41, 141 (1962).

12. Attar, A., A. H. Corcoran, and G. S. Gibson, *Am. Chem. Soc., Div. of Fuel Chem. Preprints*, 21, 106 (1976).

13. Ruch, R. R., S. J. Russell, R. Malhotra, J. D. Steele, S. B. Bhagwat, G. B. Dreher, R. A. Cahill, J. K. Frost, R. D. Harvey, and J. F. Ashby, "Determination of Valuable Metals in Liquefaction Process Residues," Illinois State Geological Survey, Final Report under U.S. DOE Contract EY-76-C-21-8004 (1979).

14. Ahmed, S. M., and B.J.P. Whalley, "The Oxygen-Flask Method of Determining Total Sulfur in Coal," *Analytical Methods for Coal and Coal Products*, Vol. 1, pp. 263-278, C. Karr, [ed.], Academic Press, Inc., New York (1978).

15. Furman, N. L., *Standard Methods of Chemical Analysis*, 6th ed., Vol., 1, pp. 1032-1035, Van Nostrand, Princeton, NJ (1962).

16. Thomas, J., and H. H. Damberger, "Internal Surface Area, Moisture Content, and Porosity of Illinois Coals: Variations with Coal Rank," Illinois State Geological Survey, Circular 493 (1976).

17. Brunauer, S., P. H. Emmett, and E. Teller, *J. Am. Chem. Soc.*, 60, 309-319 (1938).

18. Shimp, N. F., J. K. Kuhn, and R. J. Helfinstine, *Energy Sources*, 3, 93-109 (1977).

19. Raymond, R., *Microbeam Analysis*, p. 105, D. E. Newberry, [ed.], San Francisco Press, Inc., San Francisco, CA (1979).

20. Harris, L. A., T. Rose, K. DeRoos, and J. Greene, *Economic Geology*, 72, 695-697 (1977).

21. Raymond, R., and R. Gooley, *Scanning Electron Microscopy*, Vol. 1, p. 93, SEM Inc., AMF O'Hare, IL (1978).

22. Sutherland, J. K., *Fuel*, 54, 132 (1975).

23. Solomon, P. R., and A. V. Manzione, *Fuel*, 56, 393 (1977).

24. Schrauzer, G. N., V. P. Mayweg, H. W. Finck, and W. Heinrich, *J. Am. Chem. Soc.*, 88, 4604-4609 (1966).

25. Kalvnskas, J. J., G. C. Hsu, J. B. Ernest, D. F. Andress, and D. R. Feller, "Coal Desulfurization by Low Temperature Chlorinolysis," Jet Propulsion Laboratory, Final Report for Phase I for Department of Energy (1977).

KRUSE, C.W.

SHIMP, N.F.

Removal of Organic Sulfur from Coal

Exploratory studies indicate that a method for sulfur removal and liquifaction of high-sulfur coals may exist at conditions of moderate temperature and pressure.

C.R. Porter
Pentanyl Technologies, Inc.
Arvada, Co.

H.D. Kaesz,
University of California

J.L. Leto
Rocky Mountain Energy Co.

T.J. Giordano and W.R. Haas
Martin Marietta Aerospace

E. Johnson
West Virginia University

W.H. Berry, Jr.
Comsumers Power Co.

Sulfur is an undesirable constituent in coal because combustion equipment corrosion and post-combustion emissions occur. Sulfur exists in inorganic and organic compounds in coal. Although concentrations vary, roughly half of the total sulfur is inorganic and half is organic.

The inorganic sulfur in coal is principally pyritic (pyrite or marcasite, FeS_2), in concentrations of 0.5% to 5.0% of the total coal weight. It is also present as sulfate (hydrates of ferrous or ferric sulfate, e.g. $FeSO_4 \cdot 4H_2O$, $FeSO_4 \cdot 7H_2O$, $Fe_2(SO_4)_3 \cdot 12H_2O$, etc.), in concentrations of 0.01% to 1% of the total coal weight. Inorganic sulfur may exist as elemental sulfur in concentrations of a few parts per million as well. (1,2)

The principal organic sulfur functional groups in coal are (1) thiophene, (2) thiols, (3) sulfides, and (4) disulfides. (3) The chemical reactivity and relative concentrations of these organic sulfur functional groups in coal have been the subject of several recent publications. (4,5) Generally the thiophene functional groups, particularly the dibenzothiophene deriviatives, are the least and the thiol derivatives are the most reactive during desulfurization. Distribution of organic sulfur among the different groups varies in different coals, but the majority of the organic sulfur is contained in thiophenic functionalities.

Physical means of separation are currently used for removal of inorganic sulfur from coal prior to combustion. A modern plant may use a dense-media vessel for the course size fraction, and water-only cyclones or froth flotation for the fine size fraction. These and other proven mineral processing methods are currently employed to remove 40% to 90% of the inorganic sulfur in existing coal cleaning plants.

Work is being done to improve existing processes and to develop new technologies to remove sulfur from coal. A 1,200 ton per hour heavy-media plant at General Public Utilities' Homer City Generating Station is undergoing extensive study. Several new technologies, such as the Magnex Process, have been developed. (6,7)

In order to be an effective pre-combustion sulfur removal technique, coal cleaning should remove inorganic and organic sulfur. Since inorganic sulfur can be removed inexpensively with conventional technology, and since no current inexpensive technology exists for organic sulfur removal, new technologies for the removal of organic sulfur may be most significant.

Studies in support of the Magnex Process have shown that iron pentacarbonyl is an economically accessible reagent. Attention has also been focused on the reactions of iron pentacarbonyl and related low-valent metal complexes with organic sulfur coal model compounds. (8) The objective of this work is to develop a new methodology for the removal of organic sulfur from coal. A survey of the pertinent literature on how iron carbonyls react with organic sulfur compounds is summarized in the following section.

LITERATURE REVIEW

The desulfurization of organic sulfur compounds has long been known to be effected by activated metals such as Raney nickel. (9) Carbonyls and related low valent metal complexes contain metal atoms in a chemical form close to that of the metallic state. It should therefore come as no surprise that such complexes have also been shown to desulfurize organic sulfur compounds, as discussed below. The earliest report and also one of particular significance is the desulfurization of thiophene, Figure 1. (10)

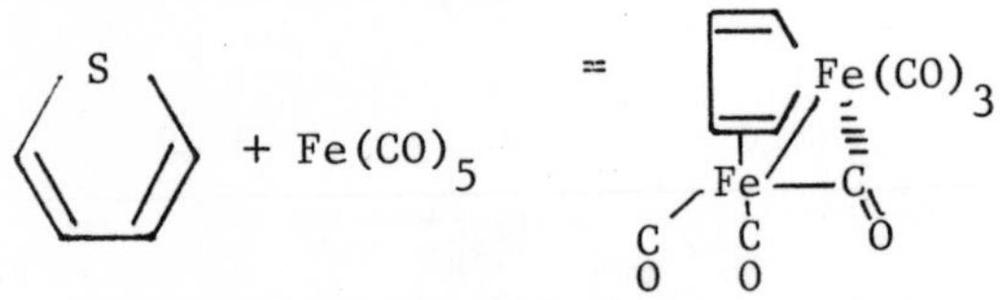

Figure 1. Desulfurization of thiophene.

This reaction is significant because it represents desulfurization of one of the most unreactive organosulfur functional groups found in coal. The lower carbonyls of iron, $Fe_2(CO)_9$ and $Fe_3(CO)_{12}$, are observed to react with organosulfur compounds as shown in Figure 1 at lower temperatures than $Fe(CO)_5$.

In the reactions of iron carbonyl complexes and certain organosulfur compounds desulfurization may not occur. With toluenedithiol, for instance, a complex belonging to the class of compounds of general formula $(RS)_2Fe_2(CO)_6$ is obtained. See Figure 2. (11)

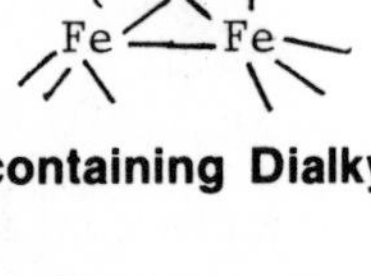

Figure 2. Sulfur containing Dialkyldithiodiironhexacarbonyl comlex.

Alper and Chan observed a similar reaction between thiobenzophenone and $Fe_2(CO)_9$ at room temperature, giving five identified products with an overall yield of 84% soluble complexes after 30 hours, Figure 3. (12)

$$Ph_2C=S + Fe_2(CO)_9 \xrightarrow[\text{30 hrs.}]{\text{benzene (rm T}^\circ)}$$

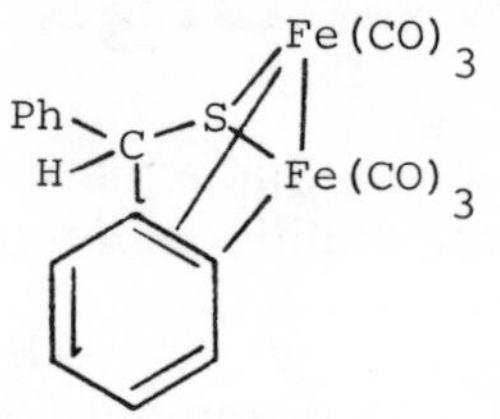

$(C_6H_6)_2CS \cdot Fe_2(CO)_6$

2, yield: 52%

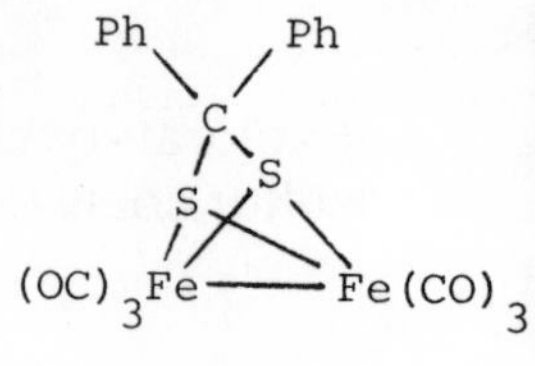

$(C_6H_6)_2CS_2 \cdot Fe_2(CO)_6$

3, yield: 5%

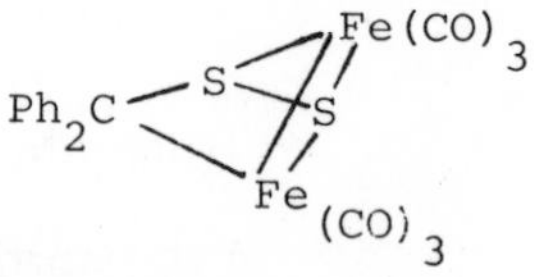

$(C_6H_6)_2CS_2 \cdot Fe_2(CO)_6$

4, yield: 9%

$S_2Fe_3(CO)_{12}$

5, yield: 18%

Figure 3. Reaction of Thiobenzophenone and Diiron Nonacarbonyl.

The chemical form in which the sulfur is combined contributes to chemical reactivity, as exemplified by the enhanced reactivity of episulfides with iron carbonyls. (13)

Nametkin et al. conducted a systemic study of the reactivity of $Fe(CO)_5$, $Fe_2(CO)_9$, and $Fe_3(CO)_{12}$ with sulfur, hydrogen sulfide, and a variety of organosulfur compounds. (14) In this work, it was found that elemental sulfur reacts with any of the iron carbonyls in benzene solution at 80°C in 1 to 2 hours, to give almost exclusively FeS. When combined with H_2S in benzene solutions at 50-55°C, $Fe_3(CO)_{12}$ gives principally $S_2Fe_2(CO)_6$. The relative reactivities of a series of alkyl organosulfide compounds with the above iron carbonyl compounds in benzene solutions at approximately 80°C were investigated, and it was indicated that $Fe_3(CO)_{12}$ is by far the most reactive, followed by $Fe_2(CO)_9$; $Fe(CO)_5$ is the least reactive. As to the reactivity of the organosulfur compounds, the alkylthiols (RSH) are the most reactive, followed by the dialkyldisulfides (RSSR); the dialkylsulfides (RSR) are the least reactive.

Another important aspect of the Nametkin work was thermal stability studies of the $(RS)_2Fe_2(CO)_6$ complexes. These show the compounds to have a temperature stability in the range 73°C (R=CH$_3$) to 129°C (R-Ph).

The Nametkin study has served as the basis of one process for desulfurization of oil and petroleum products. (15,16) This process involves treatment with iron carbonyl followed by heating, reaction with platinum-containing salts, and filtration or selective solvent extractions to remove the iron carbonyl organosulfur complexes. This work was recently reported on at a Symposium of the A.C.S. (17)

In parallel but independent studies, Hsu reacted coal with iron pentacarbonyl in an organic solvent, to remove the organic bound sulfur. (18) The reaction of Kaesz et al. (10) (Figure 1) was cited by Hsu to explain the desulfurization in which 65% to 80% organic sulfur removal is claimed. (19)

Recently, Kang et al. (20) have called attention to earlier observations that "$Fe(CO)_5$ is a very poor catalyst for the hydroformylation reaction when hydrogen is used, whereas, under Reppe's conditions with CO and H_2O, it is an active catalyst at relatively mild temperatures and pressures." More recently, King, King, and Yang reported on the rate of hydrogen production by an iron carbonyl catalyzed water gas shift reaction. In their article, several related reports were cited. (21) The proposed mechanism for the reduction of water is given in the four-step cycle shown in Figure 4.

a. $Fe(CO)_5 + OH^- \rightarrow HFe(CO)_4^- + CO_2$

b. $HFe(CO)_4^- + H_2O \rightleftharpoons H_2Fe(CO)_4 + OH^-$

c. $H_2Fe(CO)_4 \rightarrow Fe(CO)_4 + H_2$

d. $Fe(CO)_4 + CO \rightarrow Fe(CO)_5$

$CO + H_2O \rightarrow H_2 + CO_2$

Figure 4. Proposed iron carbonyl catalyzed water gas shift reaction.

The hydridotetracarbonyl ferrate anion ($HFe(CO)_4^-$) produced in the first step of the water gas shift sequence is known to be a reducing agent. The iron carbonyl catalyzed water gas shift reaction mixture generates this and possibly other related reactive intermediates. Alper has shown that solutions containing the salts of hydridotetracarbonyl ferrate anion hydrodesulfurize aliphatic or aromatic thioketones. (22) Based on this information, initial experiments were carried out at Martin Marietta Aerospace in Denver on the reactions of iron carbonyls and coal,

and of iron carbonyls and organosulfur coal model compounds. Subsequent studies were performed by Kaesz and his students at UCLA and by others.

EXPERIMENTAL STUDIES

An experimental study was initiated at Martin Marietta to evaluate the reactions of iron carbonyls with organic sulfur compounds. The objective of this study was to develop a scheme for the removal of organic sulfur from coal. Thiophenol, benzothiophene and dibenzothiophene were reacted with a solvent containing iron pentacarbonyl. These aryl organosulfur compounds reacted with iron pentacarbonyl at $75°C$ to $80°C$ and atmospheric pressure to yield iron carbonyl aryl complexes. (23) This study showed limited desulfurization of the aryl organosulfur compounds. Similarily, tests on an Ohio coal produced limit desulfurization.

It was noted in these reactions that the reactivity of the iron pentacarbonyl was increased by bubbling a nitrogen or hydrogen purge gas through the solution. A highly colored green solution formed, which indicates the presences of lower carbonyls. This procedure produced increased yields of aryl organosulfur complexes. The off gases monitored with a GC/MS during gas purging showed traces of carbonylsulfide. Although these tests showed that iron carbonyls reacted with organosulfur compounds, and demonstrated the use of a gas purge to increase reactivity, significant desulfurization was not achieved.

The results of three coal desulfurization tests are summarized in Table 1. In test NTG A-2, 100 grams of minus 14-mesh coal were reacted with a solution containing iron pentacarbonyl. This experiment was carried out in a 200 ml methanol slurry, using gentle mixing and a nitrogen gas purge at reflux temperature. After the reaction was complete, the coal was separated from the solution by filtration. Analysis showed that reaction NTG A-2 caused no significant change in the forms of sulfur or total sulfur content. In the second test, NTG 437, 100 grams of minus 200-mesh coal were reacted with a solution containing iron pentacarbonyl, with intense stirring and a nitrogen gas purge at reflux temperature. The filtered coal from test NTG 437 showed a significant reduction in the organic sulfur content and a reduction in total sulfur content. In the third test, 100 grams of minus 200-mesh coal were treated under conditions similar to the water gas shift reaction. Analysis of the coal residue from test NTG 438 showed that pyritic sulfur was apparently reduced or altered; the sulfate sulfur content was apparently increased; the organic sulfur content was very significantly altered: and the total sulfur content was reduced.

TABLE 1   ORGANIC SULFUR REMOVAL TEST RESULTS

Sample Desc.	Analytic Results[1,2]			
	Pyritic Sulfur %	Sulfate[3] Sulfur %	Organic Sulfur %	Total Sulfur %
Feed Coal[4]	0.72	0.53	1.03	2.29
NTG A-2	0.81	0.53	0.96	2.30
NTG 437	0.77	0.41	0.45	1.63
NTG 438	0.52	0.91	0.13	1.56

1) Analysis by CTE, Denver, Colorado.
2) Results on a dry basis.
3) The sulfate analysis of this coal was unusually high and was verified by three labs.
4) The feed coal was from the Ohio No. 6 Seam.

The conditions and results of three coal liquefaction tests from subsequent studies are summarized in Table 2. These tests were done in a stirred tank reactor under conditions that promote the water gas shift reaction, as described previously. Coal, solvent, catalyst and carbon monoxide were charged into the reactor, and the system was heated to the reaction temperature. After termination of the run, the products were removed from the reactor and analyzed. The analytical results summarized in Table 2 indicate that substantial conversion to THF solubles was obtained under conditions of moderate temperature and pressure. The total sulfur content of the carbonaceous material apparently was substantially reduced under the reaction conditions. The product from Test C was a free flowing liquid at room temperature.

The results of a fourth liquefaction test are summarized in Table 3. Test D was done in the conventional manner described above, but at $230^{\circ}C$, at 81.6 atm(1200 psig), and for 2 hours. The hydrogen to carbon ratios of the THF soluble fraction and the THF soluble/pentane insoluble fraction were significantly increased during processing. Also the pentant solubles or oils represented approximately 74% of the THF soluble fraction.

TABLE 2   COAL LIQUEFACTION TESTS

Sample Description	Conditions		
	Temp $^{\circ}C$	Pressure Atm(psig)	Time Hour
Feed Coal[1]	--	--	--
Test A	140	37.4 (550)	2
Test B	225	37.4 (550)	2
Test C	400	176.9(2600)	2

Sample Description	Results	
	Conversion % THF Soluble	% Total Sulfur THF Soluble Frac
Feed Coal[1]	0	--
Test A	6.9	Trace
Test B	30.8	Not Detected
Test C	93.3	0.056

Note: 1) The feed coal was from the Ohio No. 6 Seam.
2) The conversion is determined as MAF yield as follows:

$$\text{MAF Yield} = 100 - \frac{\dfrac{\text{Wt of Organics in THF Insolubles}}{\text{Wt Ash in the Insolubles}}}{\dfrac{\% \text{ Ash in Coal}}{100 - \% \text{ Ash in Coal}}} \times 100$$

TABLE 3   CONVERSION PRODUCTS RESULTS

Sample Description	Conversion		
	Pentane Soluble %	Benzene Soluble %	THF Soluble %
Feed Coal[1]	--	--	0
Test D	21.31	4.10	28.82

Sample Description	H/C Atomic Ratio[2]	
	THFS[3]	THFSPI[4]
Test D	1.53	0.89

Note: 1) The feed coal was from the Ohio No. 6 Seam.
2) The H/C atomic ratio of the feed coal is 0.84.
3) THFS means THF solubles.
4) THFSPI means THF soluble pentane insoluble.

CONCLUSIONS

Literature on these topics indicates that iron carbonyls react with organic sulfur functional groups and can result in desulfurization. Our studies confirm that the lower carbonyls (i.e. $Fe_2(CO)_9$ and $Fe_3(CO)_{12}$ and the hydridotetracarbonyl anion are more reactive with organic sulfur than $Fe(CO)_5$. Reaction of iron carbonyls with organosulfur compounds to produce desulfurization can be achieved under mild conditions ($60^\circ C$ to $80^\circ C$ and atmpspheric pressure). Exploratory studies have indicated that a method for sulfur removal and liquefaction of high sulfur coals may exist at conditions of moderate temperature and pressure.

FUTURE WORK

Programs are in progress to evaluate desulfurization and hydrogenation of other carbonaceous feeds. Tests are being done on model compounds to delineate desulfurization and hydrogenation mechanisms. Optimization studies have been initiated in preparation of pilot plant activities.

ACKNOWLEDGMENT

The authors are grateful to Consumers Power Company of Michigan and the State of Ohio Department of Energy for their cooperation and encouragement of this study. Portions of the work reported were funded by Consumers Power Company and the State of Ohio Department of Energy. The authors also acknowledge TechSearch for its contribution to this work, in performing the technical literature search and document retrieval.

LITERATURE CITED

1. Attar, Amir, Hydrocarbon Processing, 58 (1), p. 175 (1979).

2. Gluskoter, H. J., Coal Preparation, p. 1--44, Seeley W. Mudd Series, AIME, (1968).

3. Gluskoter, H. J., Coal Preparation, p. 1—45, Seeley W. Mudd Series, AIME, (1968).

4. Attar, Amir, Coal Processing Technology, IV (A CEP Technical Manual), p. 26-34, AIChE, New York (1978).

5. Meyers, Robert A. and Walter D. Hart, American Chemical Society. Division of Petroleum Chemistry. Preprints. ACS Houston Meeting, March 23-28, 1980, Volume 25, No. 2, p. 258-262 (1980).

6. Porter, C. R. and D. N. Goens, Mining Engineering, 31 (2), p. 175-180 (1979).

7. Porter, C. R., Coal Preparation and Utilization Symposium, Coal Conference and Expo, Louisville, Kentucky, October 23-25, 1979, p. 112, NcGraw-Hill, New York (1979).

8. Porter, C. R., "Organic Sulfur Removal from Ohio Coal," Final Report. State of Ohio, Department of Energy, Contract No. 80-23, (1980).

9. Challenger, F., Aspects of the Organic Chemistry of Sulfur, p. 90, Butterworths, London (1959).

10. Kaesz, H. D., et al., JACS 82, p. 4749- (1960).

11. King, R. B., JACS 85, p. 1584- (1963).

12. Alper, Howard and Albert S. K. Chan, JACS 95, p. 4905-4913 (1973).

13. King, R. B., Inorganic Chemistry, 2, p. 326-7 (1963).

14. Nametkin, N. S. and V. D. Tyurin, M. A. Kukina, Journal of Organometallic Chemistry, 149, p. 355-370 (1978).

15. Tyurin, V. D. and S. P. Gubin, N. S. Nametkin, "The New Methods of Purification and Stabilisation of the Oil and Oil Refinery Products by the low Valent Transition Metal Compounds," Proceedings of the World Petroleum Congress, 5th p. 217-222 (1975)

16. Nametkin, N. S., et al., U. S. patent 3,996,130 (December 7, 1976).

17. Nametkin, N. S. and V. D. Tyurin, American Chemical Society. Division of Petroleum Chemistry. Preprints. ACS Houston Meeting, March 23-28, 1980. Volume 25, No. 2, p. 229-241 (1980).

18. Hsu, George, C., U. S. patent 4,146,367, (March 27, 1979).

19. Pentanyl Technologies, Inc., 12191 Ralston Road, Arvada, Colorado 80004, is the owner of an exclusive license from NASA with respect to U. S. patent 4,146,367 for an invention entitled, "Coal Desulfurization," issued on March 27, 1979. While Pentanyl is authorized to sublicense such inventions, Pentanyl is not an agent or employee of NASA, nor is Pentanyl affiliated with NASA in any other respect.

20. Kang, HiChun, et al., JACS, 99 (25) p. 8323-8325 (1977).

21. King, Allen D. Jr., R. B. King and D. B. Yang, JACS 102 (3) p. 1028-1032 (1980).

22. Alper, Howard, J. Organic Chem., 40 (18) p. 2694 (1975).

23. Haas, W. R. and Thomas J. Giordano, "Organic Sulfur Removal Using Organo-metallic Chemistry," presented at the Coal Preparation Workshop, Ohio University, Athens, Ohio, May 15, 1980.

PORTER, C.R.

KAESZ, H.D.

GIORDANO, T.J.

HAAS, W.R.

BERRY, Jr., W.H.

Coal Desulfurization by Chlorinolysis--Phases II

Proximate analyses indicate a 29% decrease in volatile matter which was attributed to a polymerization of the coal during the dechlorination stage.

J.J. Kalvinskas and N. Rohatgi
California Institute of Technology
Pasadena, Ca.

Coal desulfurization by chlorinolysis was conducted by the Jet Propulsion Laboratory, California Institute of Technology, under a DOE Interagency Agreement No. ET-77-I-01-9060 with NASA for Phase II from October 1, 1978 through September 30, 1979.

An engineering scale batch reactor system was designed and constructed of acid-resistant brick for testing of coals at 2 kg of coal per batch. Operating conditions were: coal mesh size, 100 × 200: methylchloroform or water solvent medium at solvent/coal of 2/1; gaseous chlorine injection at a nominal 10 SCFH (0.28 sm^3/h) through $1/32$-in. (0.79-mm) diameter holes in a teflon diffuser; turbine agitator at 565 RPM; operating temperatures of 60 to 130°C (333-403°K); operating pressures of 0 to 60 psig (101-414 kPa); total reaction times of 45 and 90 minutes.

. The five bituminous coals that were tested in Phase II were: PSOC 276, Ohio No. 8, HVA, 3.87 wt.% sulfur; PSOC 282, Illinois No. 6, HVB, 1.59 wt.% sulfur; PSOC 219, Kentucky No. 4, HVB, 2.15 wt.% sulfur; PSOC 026, Illinois No. 6, HVC, 3.45 wt.% sulfur; and Island Creek Coal, Union County, Kentucky, 3.72 wt.% sulfur.

Normal operating procedures for the coal included: subjecting the coal to chlorination for 45/90 minutes while obtaining grab samples from the reactor for analysis at 15-minute intervals; providing a steam distillation of the methylchloroform from the chlorinator with the addition of water; providing a batch vacuum filtration and spray wash of the coal slurry; drying of the coal and dechlorination of the chlorinated coal at 400°C (673°K), 60 minutes, atmospheric pressure and nitrogen purge at 30 SCFH (0.85 sm^3/h).

A summary table of operating conditions and sulfur forms removal data is included, Table 1. Individual runs were grouped by coal type, solvent, mesh size, and chlorination stages (1). Coal analyses were conducted by the Colorado School of Mines Research Institute. A total of 44 individual runs were conducted with 15 runs on coal PSOC 276, 19 runs on coal PSOC 282, 2 runs on PSOC 219, 3 runs on PSOC 026 and 5 runs on Island Creek Coal.

Inspection of the individual runs with respect to temperature, pressure, and chlorine flow rates for the respective coals did not show any noticeable correlation with the sulfur forms data. The organic sulfur data obtained with 15-, 30-, 45- and 90-minute grab samples and methylchloroform show in some instances exceedingly high organic sulfur removals. Comparison with the fully processed bulk and dechlorinated organic sulfur removals. Comparison with the fully processed bulk and dechlorinated organic sulfur values does not substantiate these high removals.

The apparent problem is that organic sulfur determinations are obtained by difference between the pyritic sulfur, total sulfur and sulfate sulfur. With the methylchloroform grab samples, the coal appears to retain a large fraction of the iron released by pyritic sulfur; with ASTM procedure reliance on the iron determination, there is an artificially high pyritic sulfur value indicated. When the coal sample is washed extensively as in the fully processed bulk sample or when water serves as the solvent, this analytical problem is overcome. The problem with an incorrect pyritic sulfur determination can be corrected by analyzing for sulfur directly by barium sulfate precipitation instead of relying on the iron determination.

Analysis of the sulfur form data, taking into account the problem with organic and pyritic sulfur determinations in the methylchloroform grab-sample data, indicates that no substantial differences exist between runs made with methylchloroform and water in terms of sulfur forms and total sulfur removal except for PSOC 276 coal that showed an apparent higher total sulfur removal with methylchloroform than with water. (Later data have not substantiated this difference.) Also, coal PSOC 276 showed no organic sulfur removal, whereas the other four bituminous showed some, albeit low values of organic sulfur removal. Dechlorination of the coals also showed some increase in desulfurization, 5 to 10%, over that achieved in the chlorination stage. Maximum total sulfur removals for the five coals were in the range of 50 to 66%. Coal of particle size (16 × 100 mesh) relative to 100 × 200 mesh particles showed a fall-off in desulfurization for PSOC 282 coal. A second stage of chlorination provided some improvement in desulfurization conditions, if the first-stage desulfurization conditions were less than optimum.

Proximate and ultimate analyses of the five bituminous coals were carried out on both the raw coals and desulfurized coals. The analyses average for the five coals are reported in Table 2. Proximate analyses indicate a 29% decrease in volatile matter which was attributed to a cross-linked or polymerization of the coal during the dechlorination stage, since there was a corresponding increase in fixed carbon values of 24% between the raw and chlorinolysis processed coal. Ash levels were also reduced by an average of 24%. Heating values also dropped by 5.4% as a result of hydrogen loss in the dechlorination stage as HCl.

Ultimate analyses showed a 3.7% increase in carbon as a result of other coal element losses. Hydrogen was reduced by 36.6% as a result of the dechlorination process. Nitrogen was reduced slightly by 4.1%. Oxygen values increased by 20%, probably as a result of organic sulfur oxidation to sulfoxides and sulfones. Chlorine, increased from 0.23 to 2.18 wt.% for the raw vs. processed coals. The residual chlorine value of 2.18% after dechlorination was unusuallly high. Normally the coal chlorine content after dechlorination at 400°C (673°K) for 60 minutes is less than 0.5 wt.%. This high average residual value of chlorine in this case is attributed to the fact that some of the coals analyzed were subjected to two and three stages of chlorination which contributed to a higher than normal addition of chlorine to the coal structure. Total sulfur was reduced by 55% as an average for the five coals.

Ash elemental analysis was conducted for the five raw and chlorinolysis treated coals. Reduction of the ash elements were as follows: Si (-16%), Fe (59%), Al (0%), Ca (71%), K (10%), Ti (-14%), Na (29%), Mg (0%), P (0%), Mn (0%), total ash (20%).

Trace element reductions for the processed coal showed 23% for lead, 40% for arsenic and 67% for mercury. Selenium showed some reduction for individual coals but showed no overall average reduction for the five coals.

Materials balances conducted across the chlorinated, solvent recovery, coal filtration-wash stages showed an accounting as follows; raw coal, 100 ± 3%; coal organic fraction, 99 ± 3%; coal ash, 114%; sulfur, 99%; chlorine, 86%; and methylchloroform, 83 ± 7%. The high ash accounting of 114% was due to the contribution of metal corrosion that occurred in the reactor, i.e. cooling coils, test coupons, agitator shaft and impeller, etc. The accompanying coal filtrate and wash water analyses showed that the principal components in solution were: Cl^- (23-124 g/L),

$SO_4^=$ (1.5-11.1 g/L), and Fe^{+++} (1.8-12.3 g/L). Elements removed in trace quantity in solution were: Ca^{++} (.05-.43 g/L). Na^+ (.03-0.20 g/L), Al^{+++} (.03-0.93 g/L), Mg^{++} (.02-0.11 g/L), and K^+ (.01-.04 g/L). Combined filtrate and wash water to coal ratios ranged from 3.4 to 15.3, accounting for the large concentration range of the coal elements in solution between given runs.

Coal dechlorinations were conducted at 400°C (673°K) for 60 to 95 minutes using a nitrogen purge rate of 30 SCFH (0.85 sm³/h) with an average coal feed to the dechlorinator of 1078-1372 grams and an average coal chlorine composition of 6.3-25 wt.%. Chlorine removals for the five coals averaged 87 to 93%. Product coal recovery was 87 to 91%. Relatively large mechanical losses were observed in feeding and removing the coal from the dechlorinator, accounting for the majority of the losses. Volatile losses of oils and tars were observed but accounted for only a few percent of the coal feed to the dechlorinator. Subsequent improvements in coal handling to the dechlorinator have substantially minimized the mechanical handling losses.

In addition to the batch reactor program, a minipilot plant was designed and constructed to demonstrate continuous operation of the chlorinator, horizontal belt vacuum filter and spray wash, and dechlorinator. The chlorinator was designed for operation with either methylchloroform or water solvents at temperatures of 20 to 150°C (293-423°K), 0 to 100 psig (101-780 kPa) and nominal feed rates of coal of 2 kg/h at a solvent/coal rate of 2/1, with nominal retention times of 60 minutes in the chlorinator. The dechlorinator was designed to operate at a nominal feed rate of 2 kg/h of coal at temperatures of 400°C (673°K) for 30-60 minute retention time using rotary flights to agitate and move the coal feed through the reactor. Design details of the equipment are included in the Phase I final report (1). Operation of the minipilot plant is scheduled for Phase III.

Literature cited

1. Kalvinskas, J.J., *et al., Final Report Coal Desulfurization by Low Temperature Chlorinolysis Phase II*, JPL Publication 80-15 (January 15, 1980).

ROHATGI, N.

KALVINSKAS, J.J.

Table 1. Coal desulfurization data - batch reactor.

Run Nos.	Solvent	Temp. (°C)	Pressure (psig)	Rate (SCFH)	Total (kg)	15 ORG	15 PYR	15 TOT	30 ORG	30 PYR	30 TOT	45 ORG	45 PYR	45 TOT	90 ORG	90 PYR	90 TOT	Bulk[c] ORG	Bulk[c] PYR	Bulk[c] TOT	Dechlorinated[a] ORG	Dechlorinated[a] PYR	Dechlorinated[a] TOT
		Operating Conditions*		Chlorine Feed		Time (minutes)/Sulfur Removal (%)																	
Coal PSOC 276																							
1,3,7,10	Methylchloroform	65	1-50	10-19	0.68-1.32	-12	54	35	-11	73	48	-8	81	55	-	-	-	-10	83	55	-4	87	60
4,6,8,9,12,14,16	Water	65-128	1-47	6-22	0.75-1.47	-9	38	25	-14	53	34	-13	62	40	-	-	-	-16	65	40	-8	64	43
Coal PSOC 282																							
19,21,23	Methylchloroform	62-85	0-45	16-22	1.1-1.51	10	38	25	36	37	36	58	23	39	82	10	37	13	65	43	11	81	50
15,18,22,35	Water	70-90	0-57	10-13	0.80-1.66	-4	31	20	0	47	29	2	53	33	19	56	43	11	56	38	10	78	49
Coal PSOC 282 - COARSE MESH - 1/8-inch x 100 mesh																							
27,34	Methylchloroform	56-84	5-36	8-10	1.05-1.37	0	28	15	12	34	24	30	28	28	38	26	34	7	53	34	-	-	-
26,33	Water	63-71	1-37	10-12	1.33-1.56	3	17	12	0	28	17	-3	40	23	9	52	35	4	52	33	-	-	-
Coal PSOC 282 - TWO-STAGE DESULFURIZATION (Methylchloroform Solvent)																							
38-Stage 1	Methylchloroform	64-92	0-46	15	1.00	31	29	29	37	27	36	-	-	-	-	-	-	0	71	40	-	-	-
39-Stage 2	"	61-91	0-46	26	1.77	24	61	47	51	43	51	89	13	53	-	-	-	7	94	55	-	-	-
Coal PSOC 282 - TWO-STAGE DESULFURIZATION (Water Solvent)																							
36-Stage 1	Water	94-97	1-46	15	1.03	-4	36	23	-6	51	29	-	-	-	-	-	-	-4	61	34	-	-	-
37-Stage 2	"	84-93	0-44	10	0.69	1	70	40	1	71	42	4	75	44	-	-	-	3	78	45	-	-	-
Coal PSOC 219																							
28	Methylchloroform	59-88	3-38	10	1.40	31	36	46	37	40	49	62	19	55	100	-8	55	19	75	61	-	-	-
29	Water	62-70	6-34	11	1.43	-2	47	38	-6	56	40	1	73	49	-6	81	53	0	86	58	15	93	66
Coal PSOC 026																							
31	Methylchloroform	59-82	6-50	10	1.38	10	59	39	17	68	46	20	58	48	46	41	50	3	91	52	-	-	-
30,32	Water	62-92	4-37	9-12	1.26-1.62	-5	45	31	-5	70	40	-4	81	44	7	89	50	3	92	50	15	95	59
Island Creek Coal - TWO-STAGE DESULFURIZATION (Methylchloroform Solvent)																							
43-Stage 1	Methylchloroform	61-88	0-47	21	1.38	2	36	19	10	63	38	-	-	-	-	-	-	9	83	47	-	-	-
44-Stage 2	"	66-88	3-45	21	1.38	25	81	52	57	54	57	67	47	58	-	-	-	18	95	57	-	-	-
Island Creek Coal - THREE-STAGE DESULFURIZATION (Water Solvent)																							
40-Stage 1	Water	88-97	0-46	12	0.78	-8	33	15	0	38	21	-	-	-	-	-	-	5	54	29	-	-	-
41-Stage 2	"	79-93	0-50	10	0.68	-1	67	34	-1	86	44	5	89	45	-	-	-	4	87	46	-	-	-
42-Stage 3	"	94-98	10-47	12	0.79	1	76	40	8	82	44	4	87	46	-	-	-	3	86	45	15	90	53

*Conditions: Coal, 2 kilograms 100 x 200 mesh; solvent/coal at 2/1; direct steam preheat, agitator speed at 275 RPM for Runs 1-5 and 656 RPM for runs 6-44; chlorine injection for runs 1-7, 1/4-inch tubing, for runs 8-19, fritted glass diffuser, runs 20-44, 1/4 x 1/2-inch teflon tubing drileld with 1/74 to 1/16 holes, nominal size 1/32-inch.
a - Dechlorinated bulk sample for times of 45 or 90 minutes, whichever is maximum time.
b - Coal size - 1/8-inch to +100 mesh
c - Completely processed coal

Table 2. Proximate and ultimate analysis.*

	Raw Coal (wt.%)	Chlorinolysis Coal*** (wt.%)	Change (%)
Volatile Matter	36.7	25.9	-29
Ash	9.02	6.73	-24
Fixed Carbon	54.3	67.4	+24.6
Heating Value, Btu/lb.	12,949 (30.1 MJ/kg)	12,254 (28.5 MJ/kg)	-5.4
Carbon	72.1	74.6	+3.7
Hydrogen	5.32	3.40	-36.6
Nitrogen	1.52	1.44	-4.1
Oxygen (by diff.)	8.88	9.83	+20.1
Chlorine	0.23	2.18**	+1067**
Sulfur	2.96	1.33	-55

*   Average of PSOC 276, 282, 219, 026, Island Creek Coals.
**   Dechlorination at 400°C (673°K), 60 minutes. Later results show typical values at less than 0.5 wt.% Cl.
*** Chlorinated, washed and dechlorinated.

Coal Conversion R & D in Western Germany

Germany has a long experience in coal conversion. At present, the Federal Republic and the State of Northrine-Westphalia are supporting large R & D efforts for coal gasification, liquifaction and combustion carried out by the coal industry.

R. Specks and A. Klusmann
Ruhrkohle AG
Essen, West Germany

1. INTRODUCTION

Those industrial countries without major resources of oil and natural gas can expect shortages with unpredictably increasing prices for those primary energies. A sharp competition in the world trade for these energies may arise earlier than expected. Since alternative energies have to be prepared, an extensive R & D program for coal conversion has been initiated in West Germany. This program includes large projects for coal gasification and liquefaction as well as advanced combustion systems and new technologies for coal fired power stations. The products to be generated from coal are substitute natural gas and synthesis gas, liquid fractions as base products for light oil and gasoline and raw materials for the chemical industry. New power generating technologies for coal are a further target. (1)

West Germany, as a country importing almost entirely the required oil and natural gas, has steadily increased the public funding for energy R & D. Before the first oil crisis in 1973, the ratio of nuclear/non-nuclear energy R & D subsidies has been 85:1. Since then, this ratio has been reduced to 2 : 1, Fig. 1. (2)

The major part of the non-nuclear subsidy is for coal conversion. As a result of the

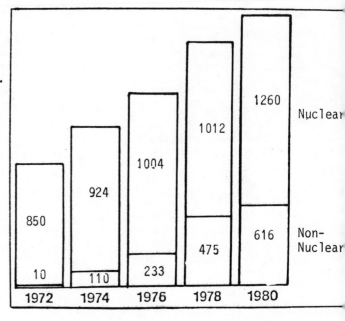

Fig. 1: Energy R & D Program (Mio DM) in West Germany

recent increase in oil prices the coal conversion programs are currently under review. The budgets being considered indicate the anticipation of a considerable increase in these expenditures in the near future, particularly if it is decided to construct large industrial complexes with public subsidies.

2. COAL GASIFICATION

Coal gasification was introduced into Germany more than fifty years ago, when the first commercial plant for town gas production went into operation in Leuna. The development and application of coal gasification continued until the end of the second world war but it was then interrupted by the expanding importation of oil and natural gas. (3)

In view of the present price increases and possible supply disturbances, the revival of coal gasification technologies has been initiated by Government programs. The objective is to prepare for the reduction of natural gas imports by the conversion of the abundant domestic coal resources using environmentally acceptable methods. Since the first oil crisis, pilot plants for various processes have been under construction or put into operation. Based on the technology of the first generation gasifiers - Winckler, Koppers-Totzek and Lurgi - the new projects habe the following objectives:

- The extension of the range of appropriate coals

- The improvement of coal conversion efficiency

- The increase of coal-throughput and gas-output per unit

- The fulfilment of the customers' specification and the environmental requirements.

The projects underway are carried out by energy companies. Fig. 2 shows the partners involved, the total costs, the project periods, the process data and the products.

TEXACO OBERHAUSEN-HOLTEN

Ruhrkohle AG and Ruhrchemie AG operate jointly the 150 t/d pilot plant based on a license from the Texaco Development Corp.. The plant is situated on the Ruhrchemie's site in Oberhausen-Holten and has been in operation for almost two years. Component tests began in January 1978 and in April 1978 the Federal Minister of Research and Technology, as sponsor of this project, performed the official start-up. (4)

Projects	Partners	Total Costs Period	Process Data	Products
Texaco Oberhausen-Holten	Ruhrkohle AG, Ruhrchemie AG	48 Mio DM 1975-1981	Entrained bed pressure gasification according to Texaco 150 t/d 40 bar 1450 C	290 000 m³/d Synthesis gas
Ruhr 100 Dorsten	Ruhrgas AG, Ruhrkohle AG/ Steag AG	145 Mio DM 1975-1984	Fixed bed pressure gasification according to Lurgi 70-170 t/d 100 bar 700-1000 C	100-235 000 m³/d Synthesis gas 40-95 000 m³/d SNG
PNP Prototype Plant Nuclear Process Heat	Bergbau-Forschung GmbH Gesellschaft für Hochtemperatur-reaktor-Technik GmbH Hochtemperatur-Reaktorbau GmbH Kernforschungsanlage Jülich GmbH Rheinische Braunkohlenwerke AG	1300 Mio DM 1975-1984	Fluidized bed gasification 1500 t hard coal/d 4000 t lignite/d Hydrogasification: 80 bar, 820-930 C Steam gasification: 40 bar, 630-800 C	1 000 000 m³/d SNG from hard coal 640 000 m³/d SNG from lignite Synthesis gas Reduction gas
KDV-Plant Lünen	Steag AG	205 Mio DM 1974-1982 —	Lurgi pressure gasification 1700 t/d 25 bar 700-1000 C	Electric Power by gas- and steam turbine 170 MW.
Shell-Koppers Hamburg-Harburg	Shell International Deutsche Shell AG Krupp-Koppers GmbH	60 Mio DM	Entrained bed gasification according to Shell: 150 t/d 30 bar 1150-1600 C	Synthesis gas Reduction gas
Saarberg/ Otto Fürstenhausen/Saar	Saarbergwerke AG Dr. C. Otto & Co., GmbH	43 Mio DM	Slag bath gasification 250 t/d 30 bar 1450-1650 C	Synthesis gas Reduction gas
High temperature Winkler-Process Frechen	Rhein. Braunkohlen-werke AG	32 Mio DM	Fluidized bed gasification 25 t/d 10 bar 870-1070 C	Synthesis gas Reduction gas
KGN-Plant Hückelhoven	PCV Gewerkschaft Sophia Jacoba	19 Mio DM	Fixed bed gasification 35 t/d 6 bar 920-1120 C	Low BTU gas Synthesis gas
VEW-Process Dortmund	Vereinigte Elektrizitätswerke Westfalen AG	18 Mio DM 72 Mio DM	Pilot Plant 24 t/d 1 bar Demonstration Plant 360 t/d 1 bar	Electric power

Fig. 2: Coal Gasification Projects

More than 5600 operation hours were achieved with several continuous test runs for periods up to 500 hours. Approx. 27000 t of different coals have been converted to 50 million m³ synthesis gas at a maximum rate of about 12000 m³/h, Fig. 3.

• Increasing the methane content of the crude gas up to 18 %.

• Using various coal qualities including low-caking coals with a grain size between 3 and 30 mm.

• Decreasing liquid by-products such as tar, oil and phenols.

• Adapting to gas composition requirements by catalytic or thermally induced gas treatment.

The project total costs of 160 million DM are sponsored by the Federal Minister of Research and Technology. (5)

Fig. 3: Plant View Oberhausen-Holten

Further development and process optimization is planned to provide reliable design data for large-scale gasification plants. The test program includes the evaluation of alternative components, alternative concepts for heat recovery and the gasification of other types of coal.

With the existing infrastucture at Ruhrchemie it was possible to construct the plant within nine months and to keep the construction costs to a minimum of approx. 10 million DM. The operating costs up to the end of 1979 will amount to approx. 26 million DM. The satisfactory results and the growing interest of the industry encourage a prolongation of the plant operation.

RUHR 100 DORSTEN

Since 1974, Ruhrgas AG, Ruhrkohle AG and Steag AG in cooperation with Lurgi Kohle und Mineralöltechnik GmbH have jointly improved the conventional Lurgi process by the development of the new RUHR 100 gasifier, to be installed at Dorsten, Fig. 4. The pilot plant has been designed for a maximum coal throughput of 170 t/d and has been completed in 1980. The plant produces synthesis gas, SNG or low-BTU gas. Particular features of the experimental program are:

Increasing the coal throughput per unit by increasing the operation pressure up to 100 bar.

Fig. 4: Gasifier Ruhr 100

PNP 500 MJ/S PROTOTYP PLANT NUCLEAR PROCESS HEAT

To improve the economic utilization of coal resources, gasification processes with a low specific consumption are required. In autothermal gasification processes the coal is not only used as raw material, but also to generate the process energy. However, the reaction energy can be supplied as process steam or heat from a gascooled high temperature reactor to save approx. 30 to 40 % of the coal. Thus, the combination of coal and nuclear enery makes possible the generation of secondary energy carriers for the heat market and the production of chemical feedstocks. The project is regarded as a long-term alternative to the autothermal process developments. (6)

The present objectives of the PNP project are

- Development and testing of hydrogasification and steam gasification of hard coal and lignite

- Development of the pebble bed HTR as a source of process heat

- Detailed lay-out of a 500 MJ/s prototype plant with an advanced HTR and gasification plant.

The total development expenses for the project period 1975-84 are estimated to be 1,300 million DM and the development of the specific hard coal gasification technology amounts to 300 million DM. The Federal Minister of Research and Technology as well as the Minister for Economics of the State of Northrine-Westphalia are responsible for a substantial part of the funding.

As future operators, Rheinische Braunkohlen-werke AG and Ruhrkohle AG established in early 1978 the joint venture company PNP to examine the technical, financial and licensing basis for the construction and operation of the 500 MJ/s prototype plant.

At present, the detailed engineering is scheduled for 1980 to 1984. A construction decision can be expected in the period 1985 to 1987. Plant operation is not likely before the early nineties and this is subject to the successful outcome of the R & D program.

COMBINED CYCLE PLANT LÜNEN

The 170 MW power plant is operated by STEAG at Lünen and represents the only advanced large-scale coal fired power plant concept in the Federal Republic. It is based on the Lurgi fixed-bed gasification process with combined cycle power station. The decision for such a plant was taken to demonstrate the following improvements compared to conventional power stations:

- A decrease of environmental impact

- An improvement in plant efficiency

- A reduction of investment costs.

Although, the low investment costs of approx. 100 million DM in the construction period between 1969 and 1972 are no longer

representative today, the plant was approx. 15 % cheaper than a conventional coal-fired plant at that time. (7)

An electrical output of 170 MW with 96 MW from an AEG steam turbine and 74 MW from a KWU gas turbine has been achieved, although the expected higher efficiency has yet to be demonstrated.

The five Lurgi gasifiers with an operational pressure of 20 bar and a coal throughput of 10 to 15 t/h can be charged with non-caking an slightly caking coal with a grain size range between 3 and 30 mm. Their operation mode and the gas cleaning system are to be improved.

The results of numerous operating tests can be summarized as follows:

- No problems with the combined cycle

- The Lurgi gasifiers need to be adapted to caking hard coal; furthermore, the re-cycling of tar has to be accomplished

- The gas treatment has to be improved to reduce the solids content of the fuel gas as well as the relatively high energy consumption.

After overcoming constructional and process start-up problems this concept is most attractive for power generation stations at sites with stringend environmental conditions.

Development work is necessary to increase coal conversion efficiency and the capacity of the gas generators. Steag AG has, therefore, decided to commission a redesigned gasifier unit in 1980 using the comprehensive experience obtained.

SHELL-KOPPERS HAMBURG-HARBURG

Shell's experience on high pressure oil gasification combined with the Krupp-Koppers know-how of construction and operation of Koppers/Totzek-gasifiers have been the basis for the joint development of the Shell-Koppers-process up to industrial scale. Following the successful operation of a 6 t/d bench-scale plant at the Shell laboratories in Amsterdam since late 1976, a pilot plant with a coal throughput of about 150 t/d has been operated at the Shell refinery complex in Hamburg-Harburg since mid 1978. (8)

The first mechanical test runs followed by short trial runs under gasification conditions have been performed. The results obtained so far have confirmed the design data. This project has not been subsidized by public funds and therefore the results are propriety data.

SAARBERG-OTTO FÜRSTENHAUSEN

Saarbergwerke AG and Dr. C. Otto & Co. are demonstrating the Rummel/Otto-system for coal pressure gasification up to 25 bar for various feed coals. Their pilot plant with a 250 t/d coal throughput was commissioned in 1978. The specific feature of the gasifier is the liquid slag bath at the reactor bottom and a system of tuyeres directed tangentially towards the bath surface, which is rotated by the impulse of the injected coal dust. The test runs are focused on the key problems of this gasifier: the controlled feedstock injection into the pressure vessel and the slag removal. With a test period of two years, the project is scheduled for four years. The costs of approx. 43 million DM are sponsored by the Minister of Research and Technology. (9)

HIGH TEMPERATURE WINKLER PROCESS FRECHEN

The conventional Winkler process has proved to be successful for highly reactive types of hard coal. With this experience, Rheinische Braunkohlenwerke AG are extending the development of the fluidized bed technology. With the results obtained from a bench-scale experiment, the basis for the pilot plant at Frechen has been established. (10)

In early 1978, the erection of the pilot plant was completed and the trial operation could start. The test program is concentrated on those points differing from the conventional Winkler process:

· Gasification under pressure up to 10 bar to raise the unit capacity and to improve the gas quantity

· Gasification at elevated temperatures up to 1100°C by adding lime, limestone or dolomite to increase the melting point of ash

· Increase of the carbon conversion rate by recycling of ungasified coal.

Up to the present, gasification tests with air have been performed up to 9 bar and with oxygen up to 5 bar. The reaction temperatures so far were limited to 850-950°C. Within the scope of the pilot plant operation scheduled to run until late 1980, gasification tests

are planned using air or oxygen/steam mixtures at pressures up to 10 bar and reaction temperatures up to 1100°C. The overall costs sponsored by the Federal Minister of Research and Technology amount to approx. 32 million DM.

KGN-PLANT HÜCKELHOVEN

Kohlegas Nordrhein GmbH (KGN), a joint venture company established by Gewerkschaft Sophia Jacoba and Projektierung Chemische Verfahrenstechnik GmbH, is developing the KGN process for fixed-bed gasification of high ash containing coals, so-called middlings. The project targets are:

· Generation of tarfree watergas or synthesis gas by cyclical operation

· Production of Low-BTU Gas for industrial boilers or gasturbines by continous pressure operation.

Sponsored by the Minister of Economics of the State of Northrine-Westphalia, the KGN gasification process has been tested first in a bench-scale plant with a coal throughput of 10 kg/h. A pilot plant with a throughput of approx. 35 t/d and a gasification pressure up to 6 bar has been completed in early 1979 at the Sophia Jacoba colliery, followed by successful start-up operation. The project cost amount to about 19 million DM. (11)

VEW PROCESS

The VEW process was initiated to generate electricity from gasified coal with minor environmental impact and higher efficiency. Special attention is given to

· The application of the combined gas and steam turbine cycle

· The combined generation of electricity and heat for various applications.

In the VEW process, coal is not completely gasified but by partial gasification an overproportional sulfur-reduction is expected to be achieved.

For more than two years, a pilot plant has been operated having a coal throughput of 1 t/h. Partial gasification of the coal with air has been tested. The results obtained so far with different coals are satisfactory. By conversion of about 50 to 60 % of the feed coal, more than 70 % of the sulfur content is separated.

On the basis of the promising test results,

the concept for a 15 t/h plant has been evaluated and the detailed engineering can be commenced in the near future. Start-up is projected for mid 1981. (12)

COAL LIQUEFACTION

Coal liquefaction was demonstrated in Germany at the beginning of this century: Bergius investigated the catalytic process, Pott and Broche the non-catalytic process. Both methods are still the worldwide basis for the present projects. Large-scale catalytic coal liquefaction started more than fifty years ago at Leuna and in the following years, coal destillates were processed to liquid fuels in six commercial plants with a total capacity of about 650,000 t gasoline per year. In 1943, the major part of the German motor fuel consumption - approx. 4 million t/a - was produced by coal liquefaction. After the end of the second world war coal liquefaction was no longer pursued in Germany for economic reasons.

The basic technological know-how of the old large projects has been used when the catalytic process became the basis of the New German Technology in the early seventies. Bench-scale experiments were started at Bergbau-Forschung GmbH and Saarbergwerke AG to establish with many process developments, new reliable data to construct modern pilot plants. (13)

The improvements in the catalytic coal liquefaction process are:

- Reduction of the process pressure and hydrogen consumption
- Increased process efficiency to reduce coal consumption
- Improved solid-liquid separation by distillation.

A survey of the new projects is given in Fig. 5.

PILOT PLANT BOTTROP

In November 1977, the decision was taken to construct a 200 t/d plant for catalytic coal hydrogenation. The plant is under construction in Bottrop close to a cokery. Waste is removed and gas, water, steam and electricity are supplied by the cokery and hydrogen is taken from a pipeline passing the site, Fig. 6. (14)

PROJECTS	PARTNERS	TOTAL COSTS PERIOD	PROCESS DATA	PRODUCTS
PILOT PLANT Bottrop	Ruhrkohle AG Veba Oel AG	300 Mio DM 1978-1983	Catalytic Hydrogenation ("German Technology") 200 t/d 300 bar 475 C	Synthetic fuels 30 t/d gasoline 70 t/d middle oil Gas 40 t/d
DEMONSTRATION PLANT Morgantown/ West Virginia	Gulf Mineral Resources Co. Ruhrkohle AG/ Steag AG	1,4 Mio $ 1975-1985	Solvent Extraction (SRC) 6000 t/d 140 bar 450-460 C	Synthetic fuels (SRC II) 1945 t/d Naphtha 440 t/d Gas 420 t/d
PILOT PLANT Baytown/Texas	DOE The Carter Oil Co. (EXXON) Japan CLDC Phillips Petroleum Co Atl. Richfield Co. Ruhrkohle AG	300 Mio $ 1974-1982	Solvent Extraction (EDS) 250 t/d 100-150 bar 450 C	Synthetic fuels 32 t/d gasoline 17 t/d middle oil 18 t/d heavy oil Gas 17 t/d
Pilot Plant Catlettsburg/ Kentucky	DOE State of Kentucky Ashland Oil Comp. EPRI Standard Oil Comp. Continental Oil Ruhrkohle AG	300 Mio $ 1976-1982	Catalytic Hydrogenation 160 (280) t/d 190 bar 450 C	Synthetic fuels 30 t/d Naphtha 50 (220) t/d middle-/heavy oil Gas 20 (27) t/d
PILOT PLANT Feldhausen/Saar	Saarbergwerke AG	30 Mio DM 1977-1980	Catalytic Hydrogenation ("German Technology") 6 t/d 300 bar 475 C	Synthetic fuels Chemical Feedstock

Fig. 5: Coal Liquefaction Projects

Fig. 6: Liquefaction Plant Bottrop

The basic engineering was carried out by Ruhrkohle AG and Veba Oel AG. Project management is also performed by both companies, the detailed engineering in cooperation with other engineering companies.

The site for the pilot plant was officially inaugurated on May 21, 1979, by the Minister of Economics of the State of Northrine-Westphalia. The mechanical completion of the whole plant and start-up operations are expected for 1981.

The investments for the 200 t/d pilot plant amount to 200 million DM. The operation costs for a three years' demonstration period are approx. 200 million DM. Furthermore, a six years' budget of 25 million DM is underway for lab-scale tests at Bergbau-Forschung GmbH to characterize the product and improve the process details. A further part of the project is the participation in the H-coal project at Catlettsburg. The overall expenses of more than 400 million DM are largely sponsored by the Minister of Economics of the State Northrine-Westphalia.

US-ENGAGEMENTS OF RUHRKOHLE AG

Ruhrkohle AG is participating in two US-projects for coal hydrogenation to get access to these most important process lines:

* The Demonstration Plant for the Gulf Solvent Refined Coal Process (SRC II)
* The Pilot Plant for the EXXON Donor Solvent Process (EDS).

GULF PLANT MORGANTOWN. Since 1975 Ruhrkohle AG with its subsidiary Steag AG has been participating in the engineering work for the 6000 t/d coal liquefaction plant planned for Morgantown, Virginia. In late 1978, the US-Department of Energy and the Federal Minister of Research and Technology agreed upon the German participation in this 1,4 million US-$ project. The German share of 25 % will be represented by Ruhrkohle AG. The Japanese Government together with industrial companies retain a further 25 % share. (15)

EXXON PLANT BAYTOWN. EXXON Research and Engineering Company and Ruhrkohle AG have signed a participation agreement for the 250 t/d coal liquefaction plant in Baytown/Texas. The other project partners are the US-Department of Energy, the Electric Power Research Institute, the Japan Coal Lique-

faction Development Co., Phillips Coal Comp. and ARCO Coal Comp. . (16)

The plant test operations began in 1980. The expenses for this project amount to more than 300 million US-$. The German share of about 10 million DM is sponsored by the Federal Minister of Research and Technology.

PILOT PLANT FELDHAUSEN

The Saarberg process for coal liquefaction is also based on the catalytic hydrogenation of coal aiming primarily at the production of distillate oils. The process has been proven in numerous bench-scale test runs. At present, Saarberg is constructing a pilot plant with a coal throughput of 6 t/d, to be in operation by the end of 1979. This project, with a total expenditure of approx. 30 million DM, is sponsored by the Federal Minister of Research and Technology and by the Minister of Economics of the State Saar. (17)

4. COAL COMBUSTION

Twenty five years ago, hard coal provided nearly 70 % of the energy in Germany. Oil replaced coal primarily in the heat market and at present hard coal contributes less than 20 % of the primary energy market production. Thus, all effort is concentrated at a time of raising oil prices to regain the original markte share. However, a greater use of coal in the future will depend upon the development of new technologies which have to meet the following requirements:

* High combustion efficiency
* Low environmental impact
* Low specific capital costs by compact design of the combustion reactors
* Utilization of low grade coals with high sulfur and ash content
* Availabily and reliability of small-scale as well as large-scale systems for industrial and utility applications.

The large field of power station improvements is not the subject of this paper which concentrates on the projects for direct coal combustion. Fluidized bed technology with atmosheric or pressurized operation is one of the new methods. (18)

In the Federal Republic of Germany, major research effort has been directed towards this development for heat generation since 1975. Fig. 7 gives a survey of the projects in West Germany, which have received substantial support from the Federal Minister of Research and Technology.

ct	Partners	Total Costs Period	Process Data		Products
ry udwig ausen	Ruhrkohle AG Thyssen Engineerng GmbH/ Standardkessel-Gesellschaft Gebr. Fasel	7 Mio DM 1977–1980	Coal Combustion at	1 bar 24 t/d .6 MW	190 t/d Steam (16 bar, 200 C) for district heating
ation ern lo.l	Ruhrkohle AG Deutsche Babcock AG	17 Mio DM 1977–1980	Coal Combustion at	1 bar 144 t/d 35 MW	1200 t/d Steam (17 bar, 450 C) for power generation
ry enau nd	Ruhrkohle AG Deutsche Babcock AG Raschka GmbH & Co., KG	19 Mio DM 1980–1985	Combustion of Colliery tailings at	1 bar 312 t/d 35 MW	720 t/d Steam (33 bar, 425 C) for industrial use
ation gen	Saarbergwerke AG	270 Mio DM 1977–1981	Coal Combustion at	1 bar 2500 t/d 600 MW	Electric Power: 220 MW
el p	Bergbau-Forschung GmbH Vereinigte Kesselwerke AG	19 Mio DM 1976–1980	Coal Combustion at	4,5 bar 96 t/d 25 MW	11,5 MW process heat 3 MW power generation

Fig. 7: Fluidized Bed Combustion Projects

KÖNIG LUDWIG, FLINGERN, GNEISENAU - PROJECTS OF RUHRKOHLE AG

The efforts of Ruhrkohle AG are focused on the development and commercialization of the atmospheric fluidized bed technology for small units. The prototype plant "KÖNIG LUDWIG" with a thermal capacity of 6 MW is a typical small boiler to supply steam to a district heating system as well as process heat for industrial apllication. For this purpose a new boiler system with a fluidized bed combustion and a coal throughput of 24 t/d is beeing installed in the plant at Recklinghausen in close cooperation with the companies Thyssen Engineering GmbH and Standardkessel. The plant went into operation in mid 1979. The costs for this project, including a two years' period of test operation will amount to approx. 7 million DM.

The second Ruhrkohle AG project "FLINGERN" is a fluidized bed installation with a thermal capacity of 35 MW to generate power. At the Stadtwerke Düsseldorf site Flingern an existing steam boiler was modified from

stoker fired operation to fluidized bed operation with a coal throughput of 144 t/d and steam raising capacity of 1200 t/d. The equipment installation was carried out by Gruppe Deutsche Babcock. After technical check-up, a two years' operation period has been started. The total expenditure for this project amounts to about 17 million DM.

Both prototype plants are intended to demonstrate the reliability of the fluidized bed system as an option for coal utilization in competition with mineral oil-derived fuels or natural gas. The simultaneous testing in two different prototype plants is advantageous, because both plants have been designed for different applications

- 6 MW KÖNIG LUDWIG plant represents a typical small boiler unit for industrial process heat generation as well as for district heating
- 35 MW FLINGERN plant demonstrates a steam raising unit for larger heating or power stations.

The operation of the two plants enables a comparison of different technical concepts with respect to the coal feed system and the boiler design, Fig. 8. (19)

Fig. 8: Fluidized Bed Power Station, Düsseldorf

The project GNEISENAU uses low grade fuels such as colliery tailings, combustible wastes or residues which stem from hard coal upgrading. Ruhrkohle AG is now planning a fluidized bed plant with a thermal capacity of 35 MW for 1980, to be located at the Gneisenau colliery in Dortmund. The project test period is scheduled up to 1985. In addition to supplying heat to the colliery and cokery complex, this project will be able to demonstrate in particular how the major problems of waste deposits can be solved. (20)

POWER STATION VÖLKLINGEN

Over a long period, Saarbergwerke AG has developed an advanced conceptual design for a prototype power station with an electrical capacity of 220 MW, to be located in Völklingen, Saar. The essential features of this power station concept can be summarized briefly as follows:

- Combined gas/steam turbine cycle
- Combination of fluidized bed technology and conventional coal dust firing system for low grade fuels
- Integration of the flue gas cleanup into the cooling tower and purification of the total stock-gas flow
- Significant decrease of NO_x and sulfur dioxide emissions
- Increase of overall plant efficiency by incorporation of the gas turbine and by improved waste heat recovery for district heating.

This advanced power station concept is expected to generate electric power by direct coal combustion more economically and more acceptably with regard to environmental concerns. When the feasibility study was completed in 1978, Saarbergwerke AG coordinated the planning work, carried out by engineering and manufacturing companies. The funds for the whole project amount to approx. 270 million DM and have been committed; the start-up date is not yet known. (21)

HANIEL BOTTROP

West Germany, the United States and the United Kingdom develop jointly the fluidized bed combustion technology in the IEA-plant Grimethorpe. Recently the Bergbau-Forschung GmbH and the Vereinigte Kesselwerke AG formed the Arbeitsgemeinschaft Wirbelschichtfeuerung to develop the pressurized fluidized bed combustion for the combined cycle process for

power production. On the site of the Steag AG power station Haniel in Bottrop a gas turbine of 3-4 MS/s is connected with a pressurized fluidized bed reactor. The most important new feature is the electrostatic precipitor for hot gas. Furthermore, the material problems of hot temperature corrosion have to be overcome. The project is sponsored by the Minister of Research and Technology and scheduled for operation for 1980. (22)

5. OUTLOOK

There is a worldwide gap between the resources and the consumption of oil, gas and coal. Therefore, oil and gas importing industrial nations have started large energy programs to save energy and to develop alternative energy technologies. West Germany imports more than 60 % of its primary energy, although there are abundant coal resources. Since the supply and the price for foreign oil and gas will be unstable factors in the future, the efforts to develop acceptable secondary energy from coal are steadily intensified. However, the new technologies require long development periods and their products are not yet economic based on German coal-prices. The difficulty for the politicians and the industry is to work out jointly the most appropriate energy program for a country which has to be flexible enough to react to world market changes, and to ensure by reasonable energy prices a positive long-term economic development trend.

Converting German coal to substitute natural gas in conventional plants will be possible from 1985 on. The estimated costs of 0,80 DM/m³ are more than twice as high as for natural gas to be contracted now for 1985. With the application of nuclear process heat after 1990 a cost reduction for SNG is expected. The natural gas price follows the oil price in Germany, therefore, with rising oil prices the present price gap between natural and substitute gas becomes narrower. Synthesis gas from coal calculated with a price of 0,24 DM/m³ is expected to be only 50 % more expensive than that generated from oil.

Coal gasoline from large conversion plants would cost 0,80 DM/l in Germany. This means 100 % more than petrochemical gasoline. Ten years ago, the coal product figures were 250 % higher, a similar relation apllies to all other liquid products from coal. The German liquefaction program aims more for consumer light oil products than for heavy power station oil, since there are only very few oilfired power stations in the country. The coal price in Germany is not expected to be as unpredictable as OPEC oil prices. Therefore, the project work for large coal conversion plants is expected to start now within the new national energy programs. First steps could be several gasification lines of 1 million t coal/a per unit and liquefaction factories with 1,5 million t coal/a throughput.

Concerning modern coal combustion, the improvement of the existing power station technology to 700 MW units operating within the environmental bounds and the demonstration of new concepts in the 200 MW class are of equal importance. The major coal consumption will remain for the generation of power. However, smaller decentral stations with fluidized bed reactors may become an important alternative in the early eighties.

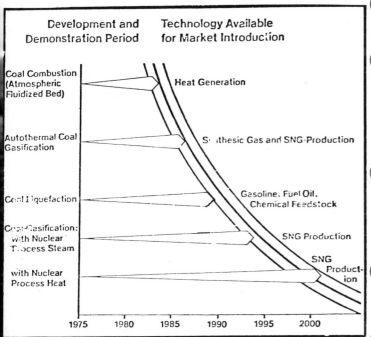

Fig. 9: Development Of Coal Conversion Technologies

A summary of these new coal conversion technologies is given in Fig. 9. With all these processes under development, coal conversion has experienced a new revival in

West Germany as in other coal countries. Large amounts of money are required and probably plenty of opportunities will arise for international cooperations to save the taxpayers' money and to provide reasonable future energy prices for the consumers.

6. REFERENCES

(1) Ziegler, A.
"THE POLICY OF THE FRG IN THE COAL RESEARCH AREA"
4th International Conference on Coal Research, October 1978, Vancover/Canada

Holighaus, R., Stoecker, H.J.
"OVERWIEV OT THE ENERGY RESEARCH AND DEVELOPMENT PROGRAM OF THE FEDERAL REPUBLIC OF GERMANY"
Symposium on Coal Refining, Edmonton, Alberta, Canada, April 20-21, 1978, Information Series 85

(2) ENERGIEPROGRAMM DER BUNDESREGIERUNG Zweite Fortschreibung vom 14.12.1977, Bonn, BMWi

(3) Specks, R.
"DIE CHANCEN DER KOHLEVERGASUNG"
Glückauf 114 (1978), 137/142

(4) Specks, R., Langhoff, J., Cornils, B.
"THE DEVELOPMENT OF A PULVERIZED COAL GASIFIER IN THE PROTOTYPE-PHASE USING THE TEXACO-SYSTEM"
Symposium on Coal Refining, Edmonton, Alberta, Canada, April 20-21, 1978, Information Series 85

(5) Röbke, G., Peyrer, H.
"DEVELOPMENT OF A NEW LURGI-GASIFIER TYPE RUHR 100"
ECE-Symposium on the Gasification and Liquefaction of Coal, 1979, Katowice/ Poland, General Report

(6) HTR-STATUSBERICHT
Kernforschungsanlage Jülich GmbH, September 1978, Jül-Spez-21, Nov. 1978, ISSN 0343-7639

(7) Meyer-Kahrweg, H.
"STAND DER ENTWICKLUNG KOMBINIERTER GAS-DAMPFTURBINEN-KRAFTWERKE MIT VORGE-SCHALTETER KOHLEDURCKVERGASUNGSANLAGE NACH DEM STEAG-LURGI-SYSTEM"
HDT-Vortragsveröffentlichungen 405 (1978), 91/97

(8) Kraaijveld, H.J.
"TECHNISCHE UND WIRTSCHAFTLICHE ASPEKTE
DES SHELL/KOPPERS-KOHLEVERGASUNGSVER_
FAHRENS"
HDT-Vortragsveröffentlichungen 405 (1978)
74/77

Völkel, H.K.
"DER SHELL-KOPPERS-PROZESS/EIN VERFAHREN
ZUR VERGASUNG VON KOHLE UNTER DRUCK"
Energie 30 (1978), No. 6, 196/198

(9) Rossbach, M., Küffner, P., Füssmann, G.,
Grams, W.,
"DRUCKVERGASUNG NACH DEM SAARBERG-OTTO-
VERFAHREN"
HDT-Vortragsveröffentlichungen 405 (1978)
78/84

(10) Franke, F.H.
"VERGASUNG VON BRAUNKOHLE ZU SYNTHESE- UND
REDUKTIONSGAS"
HDT-Vortragsveröffentlichen 405 (1978)
60/68

(11) Kaimann, W.,
"THE KGN-PROCESS FOR THE PRODUCTION OF
INDUSTRIAL GAS FROM HIGH-ASH-CONTENT
COALS (MIDDLINGS)"
ECE-Symposium on the Gasification and
Liquefaction of Coal, 1979, Katowice/
Poland, General Report

(12) Weinzierl, K.
"VEW-COAL CONVERSION PROCESS - FIRST EX-
PERIENCES AND PROSPECTS OF A NEW CONCEPT"
ECE-Symposium on the Gasification and
Liquefaction of Coal, 1979 Katowice/
Poland, General Report

(13) Strobel, R., Bönisch, U. Friedrich, F.
"ERGEBNISSE DER VERSUCHSANLAGE KOHLEÖL
DER BERGBAU-FORSCHUNG"
HDT-Vortragsveröffentlichungen 405 (1978)
42/46

(14) Langhoff, J., Wolowski, E., Esscher G.,
Hosang, H.
"FURTHER DEVELOPMENT OF GERMAN TECHNOLOGY
COAL HYDROGENATION BY RUHRKOHLE AG AND
VEBA-CHEMIE AG - THE 200 T/D-PILOT PLANT
BOTTROP"
ECE-Symposium on the Gasification and
Liquefaction of Coal, 1979, Katowice/
Poland, General Report

(15) Jackson, D.M., Schmidt, B.K.
"COMMERCIAL SCALE DEVELOPMENT OF THE
SRC II PROCESS"
5th Annual International Conference on
Commercialization of Coal Gasification,
Liquefaction and Conversion to Electri-
city, 1979, Pittsburgh, USA

(16) Epperly, W.R., Taunton, J.W.
"WXXON DONOR SOLVENT COAL LIQUEFACTION
PROCESS DEVELOPMENT"
Coal Dilemma II, American Chem. Soc.,
Industrial and Engineering Chemistry
Division, Colorado Springs, Col. Febr. 7

(17) Würfel, H.
"DAS PROJEKT KOHLEHYDRIERUNG DER SAAR-
BERGWERKE AG"
HDT-Vortragsveroffentlichungen 405 (1978
38/41

(18) Schilling, H.D.
"DIE WIRBELSCHICHTFEUERUNG - EINSATZMÖG-
LICHKEITEN FÜR DIE STROM- UND WÄRME-
ERZEUGUNG AUS KOHLE"
VDI-Bericht No. 322, 1978, 1/8, ISSN
0083-5560

(19) Stroppel, K.G., Langhoff, J.
"DEMONSTRATIONSANLAGEN KÖNIG LUDWIG UND
FLINGERN DER RUHRKOHLE AG, ESSEN"
VDI-Bericht No. 322, 1978, 37/43, ISSN
0083-5560

(20) Asche, V.
"BESEITIGUNG VON FLOTATIONSBERGEN DURCH
WIRBELSCHICHTVERBRENNUNG - PROJEKT
GNEISENAU"
VDI-Bericht No. 322, 1978, 7/12

(21) Meyer, W.
"EIN KOMBIBLOCK AUF STEINKOHLENBASIS MIT
WIRBELSCHICHTFEUERUNG"
VDI-Bericht No. 322, 1978, 45/48

(22) Schreckenberg, H., Schilling, H.D.,
Wied, E.
"STAND UND PLANUNG DER GASTURBINENANLAGE
DER ARBEITSGEMEINSCHAFT WIRBELSCHICHT-
FEUERUNG (AGW)"
VDI-Bericht No. 322, 1978, 97/101